CLEARER, CLEANER, SAFER, GREENER

CLEARER, CLEANER, SAFER, GREENER

A Blueprint for Detoxifying Your Environment

Gary Null

an OMNI Book

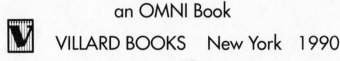 VILLARD BOOKS   New York   1990

Null, Gary.
    Clearer, cleaner, safer, greener: a blueprint for detoxifying your
environment / by Gary Null
      p.   cm.
    ISBN 0-394-58316-7
    1. Environmental health—Popular Works.   I. Title.
    RA566.N85 1990
    613—dc20   89-43467

Manufactured in the United States of America
98765432
First Edition

ACKNOWLEDGMENTS

During the past few years I've had the good fortune to have worked with a unique, brilliant, insightful, probing, and challenging mind. That mind belongs to Trudy Golobic. Trudy has been of invaluable assistance as a creative editor helping in the conceptualization and execution of this work. I feel fortunate to have a working relationship with someone who possesses such unique gifts, including patience. This book required many years of research, more than a thousand interviews, and reading thousands of scientific reports, articles, and books, all in an attempt to understand the complete picture of the relationship we have with our planet. I thought I had a complete manuscript on five separate occasions over a five-year period, but intuitively I felt much more remained to be understood, researched, and investigated. It is only in the past twenty-four months that environmental issues have received the attention they long deserved.

When I finally finished my research, Babs Lefrak, editorial director at Omni Books, believed the public was now awake to the issue. She worked skillfully and dilligently to help in the final molding of what was a vast amount of material to bring it to a final form that is much more readable and hence more palatable than my own drier and more technical manuscript. I appreciate her substantial contributions to this book.

I would also like to thank Diane Reverand for her enthusiasm and encouragement on this project.

Contents

INTRODUCTION
Where We Live and Work

Elaine hops off the subway into the fresh, crisp air outside, feeling good about herself and about life. She walks the few blocks to her office, takes a good deep breath of air, and pushes into the Fifth Avenue building. Within half an hour, Elaine looks and feels like a different person. It starts with her eyes burning, then the liveliness she felt earlier begins to fade into an overriding sense of fatigue. Coffee perks her up for a little while, but shortly afterwards her eyes grow even more irritated and she feels as though she can hardly hold herself up. At five o'clock she sighs, unable to believe that she actually made it through another day. She drags herself out of the office, down the elevator, and then, like a miracle, within a few minutes she begins to feel great again.

For years Elaine jokingly referred to her condition as an "allergic reaction to work." She had gone to see a number of doctors and a psychiatrist, and all of them more or less agreed. It was in her head, they said. No other explanation for it. And actually it was not all that uncommon. A lot of people experienced many strange fatiguelike symptoms in connection with work. Elaine's psychiatrist suggested that she was acting out not wanting to take care of herself.

Recently, Elaine got a new job. She was excited about it, but in a way was fearing it too. Her new position called for a fair degree of responsibility and thinking on her feet. What if her lethargy took over and she couldn't think, as so often happened at her old job? Fortunately, that never happened. Elaine found that in her new sunny office she felt just as wonderful as she did before getting into work. Was it merely a coincidence?

Elaine never really knew for certain what the problem was. When she heard about others who were having experiences similar to hers because of a "sick building" or indoor air pollution, it all made sense. While she felt good in her new office, she noticed that it had windows that opened, natural lighting, and was decorated with soft colors. Nobody smoked. She was far away from a photocopier, and her new employer encouraged her to get up, stretch, and take breaks, particularly when she worked intensively on her computer video-display terminal. On the other hand, she noticed that when she went to visit clients in their office, she would often start to experience the same symptoms all over again.

Studies reveal that conditions such as Elaine's are not unusual. Once scoffed at and rejected as a figment of an overly indulgent imagination, a wide variety of health problems are now being attributed by health authorities to the poor quality of the air inside our homes and office buildings. In 1985, the Environmental Protection Agency (EPA) completed a study of air pollution within the home. Volunteers from across the country took part in the five-year study, wearing monitoring devices sensitive to eleven environmental contaminants, including chloroform, trichloroethylene (TCE), and benzene. The study found that no matter where people lived—in large cities or remote country areas—levels of the toxic contaminants were much higher indoors than they were outside. "It's probable," the study concluded, "that major sources are consumer products such as paints, cleansers, mothballs, deodorants, plastics, building materials, dry-cleaned clothes, gasoline, and cigarettes."

Since the energy crunch of the 1970s, homeowners and landlords have been sealing up buildings in attempts to reduce energy costs and consumption. This energy-conscious rush to save pennies may ironically be costing us billions in lost productivity, medical expenses, and environmental quality. Although the air in many of these energy-efficient properties has become saturated with chemicals, fumes, and toxic gases, the costs of rehabilitating them may be so exorbitant that many landlords will either deny the existence of problems or indefinitely postpone much-needed modifications.

In May 1989, the EPA's long-awaited report on the quality of indoor air made its unofficial debut when Senator Frank Lautenberg, a Democrat from New Jersey, distributed bootleg copies of the EPA draft to members of a Senate subcommittee. Putting indoor air quality at the top of the list of the nation's environmental problems, the EPA report made the following findings:

•A major share of the public's exposure to air pollution is received indoors and may result in serious acute and chronic health risks. The evidence warrants an expanded effort to characterize and eliminate this exposure;

• Annual national costs of medical care resulting from major indoor air pollution [IAP] health effects are over $1 billion;

• Potential health effects of major indoor air pollutants range from itchy eyes and runny noses to chronic organ damage, and death from lung cancer and other diseases. The extent to which such effects actually occur depends on many factors, including the degree of exposure and the susceptibility of the individuals exposed;

• Few estimates are available for noncancer health impacts but many scientists believe that these may be the most common and important effects of IAP;

• Exposure to IAP in nonindustrial environments poses a significant threat to the population. We are all exposed to radon, and other pollutants that pose significant cancer risks. These include ETS (environmental tobacco smoke) and VOCs (volatile organic compounds). Radon and ETS are present in a very large number of homes, and ETS is present in office buildings. VOCs are ubiquitous in indoor environments. Additional cancer risks come from asbestos, formaldehyde, PAHs (polynuclear aromatic hydrocarbons), and pesticides in indoor air.

A building can become "sick" and indoor air contaminated by a wide variety of factors—inadequate air and poor ventilation being by far the most important. This problem first attained public notoriety in 1976 when twenty-nine people died during a Philadelphia convention from the mysterious Legionnaire's disease. Health authorities later traced the disease to improper cleaning and maintenance of the hotel's air-conditioning system. Bacterial growth was allowed to accumulate, and vaporized throughout the building by the cooling system, causing the deadly infection when inhaled. Chemist Gray Robertson, president of ACVA Atlantic, a company that cleans up sick buildings, contends that poorly maintained ventilation systems cause disease.

The bacteria causing Legionnaire's disease is commonly found in the soil, where it is present in dilute concentration and normally does not affect people. If some of the soil is disturbed and carried in air currents due to construction activity, for example, particles can be sucked into building ventilation systems. If the soil particles fall onto a source of moisture such as the cooling tower or condensing trays, the bacteria can begin to multiply at a colossal rate. As the air comes in over the coils and through the cooling tower, some of the water evaporates, carrying with it the bacteria, which are now present in massive doses and distributed through the ducts throughout the building.

Another problem is the huge amount of debris that accumulates inside cooling systems, heating units, and duct work. Any contamination in the ducts will carry through vents to the people throughout the building. Robertson has found dead birds, snakes, rats, and insects, not to mention hundreds of pounds of beer cans, food wrapping, and other filth that constantly recontaminates a building's air.

Bacterial infection from a sick building is not unusual. Dr. Alfred Munzer, a director of the American Lung Association, has examined many patients with vague respiratory problems that are quite bothersome. After interviewing them on their living and working habits, he found that the symptoms are associated with something at work.

A building-inspection division of Honeywell compiled data taken from buildings surveyed over a year and a half. The report concluded that:

•In almost two thirds of the buildings inspected, there was a measurable increase in toxic fumes, ambient heat from lights, and VDTs, and a noticeable hike in the number of workers occupying the same space.

•Some buildings infuse no fresh air into their ventilation systems.

•In 75 percent of the buildings, maintenance was largely overlooked. Clogged filters and trash in air ducts were found throughout.

While the health effects of things like tobacco smoke and asbestos are fairly well accepted by medical and health officials, the E.P.A. itself admits that it has little in the way of hard facts concerning the adverse health effects of many indoor pollutants. During air-conditioning season, many people come down with summer colds or viruses from work. Most often these are attributed to a change in temperature between the warmth outside and the cool air inside. But how many of these illnesses are in reality caused by bacterial or viral agents wafting through the air from poorly maintained cooling systems?

Quite apart from bacterial contamination, there are a large number of other substances that can accumulate and make people feel ill in tightly sealed buildings. We know a woman who started noticing a number of years ago that whenever she did photocopying for any period, she would begin to feel sleepy at first, and then experience achy and flulike symptoms. She asked a number of people if they had ever felt the same way, and they said no—it was all in her head. We now know, however, that copying machines as well as computers, new carpeting and furniture, and paint can all "outgas" fumes and vapors that can have significant health effects. Employees are now suing companies and building designers for medical conditions that result from indoor air pollution, especially when it's been learned that employers were aware of the problems. Until recently workers' compensation laws prohibited workers from suing employers over most safety and health issues. Employees are finding their way around this by demonstrating the foreknowledge of companies and building owners.

When a publishing company moved its headquarters to a new, tightly sealed building on San Francisco Bay, workers almost immediately began to complain of stinging eyes, sore throats, and various other symptoms. One editor, who had worked for the company for many years, was particularly hard hit, and was out for weeks at a time with respiratory ailments. Workers noticed that people on the second floor of the building were the hardest hit. When they took their complaints to management, employees were met with what they called a conspiracy of silence. Frustrated by employer apathy, the ailing workers took the extraordinary step of forming a union. They discovered that the source of the contamination was a gasoline pump used to refuel boats at the marina located next to the second-floor air-intake duct. The employees have not yet decided whether they will take legal action against their employer.

The phenomenon of the sick building is not an easy one to analyze. Yes, sealing up our buildings and eliminating the flow of fresh air is certainly a major factor. But the problem is also escalated through a number of synergistic effects. In cities like New York, for instance, office space has become so expensive that a greater number of people are crowded into smaller spaces. As a result, air becomes stuffy, carbon monoxide and dioxide begin to accumulate, and people begin to feel tired, depressed, and are less productive. If smoking is allowed in the office, the problem is amplified. We also have more office machinery—computers, copiers, and fax machines—all of which emit electromagnetic radiation as well as fumes and gases that may be causing many more health problems than are currently recognized.

matically over the past few years. With no place to go, substances like formaldehyde from carpets and furniture can remain trapped in an office for weeks. Even people who do not react immediately to chemical contamination may suffer immune depletion and begin to react after periods of exposure.

Further complicating identification of indoor-pollution problems is the individual nature of people's responses to environmental contaminants. Doctors are beginning to recognize that allergic-type reactions to substances in a person's environment can evoke a wide range of responses. Some people with strong immune systems, for instance, may have no allergies or sensitivities at all. Others may react violently. Dr. Sherry A. Rogers, a specialist in allergy and environmental medicine, has tracked reactions to mold in a forty-one-year-old man who experienced headaches, dizziness, and extreme fatigue, to severe eczema in another, to an eight-year-old boy diagnosed as hyperactive until molds were discovered as the cause of his erratic behavior.

Many reactions to environmental factors can cause what Dr. Rogers calls "brain fog" or "toxic brain syndrome," where the substances evoke chemical reactions within the brain. There are no blood tests or X rays to diagnose it. Physicians unfamiliar with environmental medicine usually label these patients as hypochondriacs and suggest they get psychiatric help.

Sources of Pollution

In addition to the tightening of our buildings and the growing number of chemicals in our environment, the effects of indoor pollution are compounded by the amount of time most people spend inside these days. In fact, unless you work out-of-doors or conscientiously exercise, chances are you may spend as little as one to two hours a day in the fresh air. Long-term constant exposure to toxins within the indoor environment can be the source of significant health problems. A discussion of some of the most significant sources of indoor pollution is helpful at this point.

Carbon Monoxide: The National Center for Health Statistics estimates that 2 percent, or 4.5 million people, are exposed to indoor levels of carbon monoxide that exceed the federal safety limits. About 200 people die each year from carbon-monoxide poisoning from space heaters alone. Symptoms can vary, depending on the intensity of the exposure and an individual's reaction, as with many other environmental toxins.

Many of our gas, coal, kerosene, and oil appliances—heaters, furnaces, stoves—give off significant amounts of carbon monoxide, particularly when they are not properly adjusted or maintained. Appliances don't have to be defective to be dangerous. Many portable space heaters are not vented outdoors. A Yale Medical School study found that some of these heaters can pollute the air of a typically ventilated room to a level of about 12 parts per million (ppm) of carbon monoxide. A day spent in a room with such a heater can bring blood carboxyhemoglobin levels of healthy nonsmokers almost to the halfway level known for the onset of nervous-system symptoms such as chronic fatigue and brain fog.

Radon: In January 1988, the National Academy of Sciences released a report estimating that as many as 13,000 lung-cancer deaths occur each year as a result of exposure to radon, an odorless gas emitted by underground rocks that seeps into buildings through cracks in the foundations. The effects of radon are greatly aggravated by cigarette smoking.

In September 1989, the Environmental Protection Agency (EPA)—previously criticized by many as downplaying the health risks of radon—issued its own report concluding that radon represents one of the nation's worst environmental threats, endangering hundreds of thousands of households each year. Federal authorities have issued a national public-health advisory urging Americans to test their homes for the naturally occurring radioactive gas.

Radon is an invisible gas that is produced when uranium in the soil deteriorates and gives off radioactive particles. Outside, the gas dissipates in the environment and causes few problems, but when it seeps into buildings and accumulates because of inadequate ventilation, radioactive particles can enter the lungs and cause cancer. Although the extents and risks of exposure are still not known for certain, we do know that the dangers are significant. According to a survey of 11,000 homes in seven states, the EPA found that one third of the homes tested had levels above the federal standard of four picocuries per liter of air. For people who spend three quarters of their time indoors, that's like smoking a pack of cigarettes a day. Officials also estimate that at this level of exposure over a period of 70 years, 20 percent would die of lung cancer as a result of radon exposure.

Results from over-the-counter test kits may be misleading or false because of the poor quality of laboratories doing the analysis. You can get recommendations from both state and federal environmental-protection agencies for testing and safety levels.

Formaldehyde: In 1984, the EPA designated formaldehyde for priority attention. In its report on the quality of indoor air, the agency emphasizes the need for additional federal regulations affecting formaldehyde emissions from urea-formaldehyde-pressed wood products (particleboard, plywood paneling, and medium-density fiberboard).

Formaldehyde, the active ingredient in embalming fluids, is found in a wide variety of consumer products ranging from carpets and upholstery, permanent-press fabrics, paper products, and cosmetics (some nail polishes and eye makeups can contain up to 5 percent formaldehyde). It is a primary ingredient in urea-formaldehyde foam insulation, which was banned in 1982 by the U.S. Consumer Product Safety Commission after persistent health problems were linked to its fumes. A subsequent decision by the Fifth Circuit Court of Appeals held the safety commission's ban to be invalid, and the insulation process is still available despite its known health risks.

The chemical is a potent irritant, so it's difficult to tell whether symptoms arise from its irritating side effects on the eyes, nose, and throat, or whether formaldehyde works directly on the brain.

Common symptoms include chronic eye irritations, respiratory problems, rashes, fatigue, confusion, and chronic thirst. Studies indicate that it can cause cancers of the sinuses, lungs, and liver.

Asbestos: This fibrous substance was widely used as an insulator in almost all of our buildings until its health risks were recognized in the mid-1970s. While asbestos is no longer used, major programs have been instituted by the EPA to help defray the costs of getting the substance out of our schools and public buildings.

Health risks from asbestos result from microscopic fibers that can enter the lungs and remain there as an irritant for years. These fibers are recognized as the cause of mesothelioma, one of the worst forms of lung cancer. Scientists now believe that even a single asbestos fiber is capable of causing a tumor over time.

Volatile Organic Chemicals: After World War II, organic chemicals began to be used in a wide variety of consumer products. Today they are ubiquitous. They can be found in fabric-care products, disinfectants, paint products, and furniture polishes and waxes. Fabric-care products like spot removers and dry-cleaning solutions, for instance, can contain the carcinogens benzene and toluene. Dichlorobenzene, methylene chloride, and trichloroethylenes are often used in shoe dyes, polishes, and cleaners. Their toxic effects include heart, liver, and kidney damage, bladder cancer, and respiratory disorders.

Instead of getting rid of smells, room fresheners and deodorizers add even stronger and more toxic fumes to our environment.

Many of them "deodorize" by deadening the nerve endings in our noses. Germicides and disinfectants contain phenol and cresol. Phenol is a recognized carcinogen. It is easily absorbed through the skin and can cause damage to the central nervous system, liver, kidneys, and other organs. Paint and paint removers almost all contain organic solvents such as benzene, toluene, and xylene. Benzene is considered one of the environment's most dangerous chemicals. Fumes from all of these products can be highly toxic. When such fumes are combined and trapped within an poorly ventilated space, health effects can be significant.

Cleaning Up the Office

Comfortable and aesthetic office spaces traditionally have been the domain of prestigious companies and their officers. Office buildings, with their marble and brass lobbies and spacious window suites, were designed with that in mind. But comfort and beauty in the workplace usually ends there. Small firms often do not spend much on their office surroundings. While large corporations and law firms may have luxurious offices, more often than not the support staff is crowded into cubicles lacking in privacy, as well as adequate lighting and ventilation. We now know that the quality of our indoor environment has significant effects on the health and productivity of workers. Accordingly, the first step toward a safer and healthier working environment is an attitude shift. Employers and workers alike need to recognize that things like proper lighting and ventilation are necessities, not luxuries. And changes need to be made accordingly.

Building owners must be pressured by tenants to provide more fresh air in their buildings in order to dilute the pollution to relatively harmless percentages in the ambient air. The best way to do this is, of course, to provide windows that open, so people can ventilate their own space according to their needs. For landlords who refuse to replace sealed windows with ones that open, the next-best solution is to make sure that the ventilation system is adequately cleaned and blows sufficient quantities of fresh air into the office. This is also important in situations where window offices are reserved for employees higher up in the company hierarchy, while the support staff is clustered into the center of the space. Often the air in these center workstations is intolerably close and stuffy. The lack of oxygen alone can make workers tired and hazy. Nobody knows what the cumulative health effects and long-term costs can be of things like passive cigarette smoke, carbon monoxide, and the heat and pollution from office machinery.

Lighting is another factor that is only beginning to get the attention it merits in offices and public spaces. Although people have been complaining about it for years, fluorescent lighting was ubiquitous in our commercial buildings. We now recognize that this type of illumination causes a number of health problems. First, the common, commercial fluorescent tube lacks the full spectrum of light contained in natural sunlight. This deficiency has been shown to cause mood swings, depression, and decreases in productivity. Second, fluorescent light operates with small pulses of energy that are too quick to be consciously observed, but that are nevertheless registered in the brain. The result of these constant flashings can be fatigue, eyestrain, and eventually damage to the central nervous system. Adequate natural light is the best solution. If that is not possible, full-spectrum lighting is now available, and employees need to insist that their employers install it, preferably in a nonfluorescent form.

A common problem in older buildings is asbestos located in the building's insulation and duct work. This absolutely needs to be removed, and no reputable landlord should quibble about the cost. With the current body of scientific literature concerning the carcinogenicity of asbestos, landlords who fail to remove asbestos leave themselves open to negligence suits in the future.

New carpets, furniture, paints, and building supplies commonly give off toxic fumes and gases when they are first installed. Nontoxic sealants are now available for application to carpets and upholstery to seal in most formaldehyde and other chemicals that emit fumes.

Almost all office equipment today emits varying amounts of pollutants. Photocopy machines should always be placed in well-ventilated areas. Video-display terminals (VDTs) emit large electromagnetic fields that studies are now linking to miscarriages in women who use computers intensively, and possibly central-nervous-system damage. The VDT, like the fluorescent light, also pulses and can cause considerable eye fatigue and visual damage. To decrease potential health risks, avoid staring directly into the screen whenever possible. Look away, for instance, when you are entering material. Do outlines before you sit down at your computer to decrease the time you sit in front of it, and use printouts for reading and editing instead of gazing into the screen.

Detoxifying the Home

Since everyone's home is different, the first thing you need to do is analyze your own situation and any problem you may have. For instance, do you rent or own your home? If you are an owner, are you thinking about building a new home, renovating an existing one, or do you just want to make minor health-oriented alterations? If you have a particular problem, you need to analyze it. Do you feel worse in one room than in another? Does anyone else in your house? Don't forget to take pets and houseplants into consideration. Animals may react strongly to formaldehyde because their noses are closer to carpets; leaves on houseplants may get brown if there is a toxic problem indoors. If you have pets, you need to consider whether they may be the source of a problem. If they urinate on rugs, bacteria may be accumulating, and animal danders are common allergens for many people.

You also need to consider the location of your home, since the quality of outdoor air will directly affect indoor quality. If you are planning to build and are selecting a site, there are a number of factors you will want to avoid:

•Industrial areas: Many manufacturing processes involve the use of toxic chemicals that may be dumped into rivers or released into the air. Toxic-dump sites invariably leak and contaminate groundwater in surrounding areas.

•Highways and gas stations: Both are sources of toxic gases. Carbon monoxide and ozone can cause respiratory problems. Leaking storage tanks at gas stations can contaminate groundwater.

•Sources of high emissions of electromagnetic radiation: This includes utility lines, broadcasting towers, and airport radar systems. Studies now link high-voltage power lines to cancer, particularly in children, as well as damage to the central nervous system. For this reason schools should not be situated near power lines.

In constructing or remodeling a home, you may want to consider the following factors:

•Gutters should always drain away from buildings so that they do not allow water to accumulate near foundations, walls, and in the basement. Excessive moisture can allow allergy-causing molds and fungi to proliferate.

•Ideally a house or any building should be built tightly but with good ventilation. In this way air flow can be controlled. Air filtering through walls can blow bits of fiberglass and other particulates, including asbestos in older buildings, indoors.

Indoor pollution can result from materials used in construction, furniture, and other products used inside, so it is important to inventory these factors:

•*Floorings:* Wood, tile, and other smooth surfaces are less prone to accumulating dust, molds, and other biological allergens than carpeting. Unfinished surfaces of wood, brick, and unglazed tiles, however, have pores in which dust, kitchen grease, and all sorts of dirt can accumulate. Untreated, these surfaces also begin to fade and become gray. Hard surfaces are usually treated with urethane products. These chemicals work well to seal the surfaces and give them a durable finish, but they have two major defects. First, they emit highly toxic fumes that can evoke strong reactions in some people. Second, urethaned floors need to be refinished periodically, and in order to do this properly, the old layer needs to be stripped. This can be a costly and time-consuming job, and requires the use of more toxic chemicals.

As an alternative to urethane finishes, you can now buy penetrating oils made with nontoxic ingredients that actually seep into the cellular structure of the wood or ceramics. The result is a smooth, satiny finish that can last for years. The oil sealant is usually followed by waxing, and occasional waxing and buffing is recommended for heavily used areas. One such product is the Kaldet Resin & Oil Finish manufactured by Livos Plant Chemistry. These sealants also come in colors and can be used on other woodworking to protect it from moisture.

If you are choosing tiling, remember that soft vinyl tiles outgas more than hard tiles. Some older tiles were also made with asbestos and should be removed if that is the case.

If you do have carpeting, make sure to vacuum and shampoo it frequently. Carpeting is a repository for all sorts of food particles, bacteria, molds, yeast, dust, and animal danders. Microorganisms thrive particularly when there is excess moisture in the air. And don't forget to change vacuum cleaner bags often—full bags can blow dust and particles back into your air and also cause machines to overheat and give off fumes.

Removing or replacing carpeting is a number-one priority if you are concerned about the quality of your indoor environment.

While older carpeting is probably not outgassing, even conscientious vacuuming and cleaning will not totally eliminate biological allergens.

Synthetic carpets give off a host of toxic fumes and allergy-evoking particles. Not only do almost all synthetics contain formaldehyde, but also things like ethylbenzene, toluene, styrene, benzene, and diphenol ether. Some people react immediately with violent symptoms to these chemicals. Others may not react at first, but constant exposure can lead to suppression of the immune response and a resultant sensitivity not only to the particular chemicals in the carpet, but to a number of other substances. Fibers from synthetic carpeting also deteriorate with time and contribute to household dust. When they enter the home's heating system, they can burn and give off even more fumes.

Many of the newer systems of carpet installation glue padding to a particleboard or plywood flooring. The glue may give off fumes, and the flooring is often treated with formaldehyde. Removing carpeting altogether under these circumstances may be quite expensive. One alternative is to replace synthetic carpeting with one made of wool and colored with nontoxic dyes. Careful taping and installation should be able to cut down on fumes from particleboardlike flooring. Another solution is to tape up both the padding and carpeting, carefully seal the particle-board flooring, and then build a wood floor over it.

Man-made wood floorings like particle board (also used in kitchen cabinets and some walls) contain a number of toxic chemicals that can outgas for many years. Surfaces made of these synthetic woods should either be sealed with foil under the carpeting or coated with nontoxic sealants to seal in formaldehyde and other gases.

If you suspect that tiling in your house contains asbestos, it may be best to cover it with something else. Ceramic tiles are an excellent choice for bathroom walls and floors. They can be expensive, but you can save money by buying seconds.

Walls, Doors, and Windows: Depending upon the age of your home, you may have a variety of structural problems. Older buildings used asbestos not only between walls as an insulator, but also in ceilings and walls themselves. You can recognize it in cottage-cheese-type ceilings and walls. You will also want to look at your ducts to see whether they have been insulated with the fiber.

Newer structures, on the other hand, have the problem of formaldehyde vapors escaping from plywood and other manmade wood products as well as from the glue that holds them together.

Most paints, stains, and paint removers are made with organic solvents like petroleum ethers or aromatic hydrocarbons. The ethers are highly flammable, while the aromatics include toxics such as benzene, xylene, and toluene. Older paints also contain heavy metals like lead. Many paints, both old and new, can enter easily into the bloodstream upon contact, causing liver damage. The petroleum by-products also outgas significantly. Today, there are a number of paints available for sensitive people. Unless your walls really need repainting, you also might want to consider giving them a thorough wash down with soap and water. This is much less toxic and expensive than repainting.

Condensation on walls is an indicator that the moisture level in the building is too high. Excessive moisture can lead to proliferation of molds, mildew, fungi, and other biological allergens.

In the basement, you need to check to make sure that there are no cracks in the foundation through which radon can leak into the building.

Ventilation and Heating: Adequate ventilation in every part of the house is essential for a healthy indoor environment. In the kitchen, a gas stove can be very polluting if the room is not ventilated either by a window or a good fan. Make sure that your fan vents to the outside of the building and does not simply blow air back into the room. In the bathroom, you need ventilation to remove excess moisture from the air. Even with a good bathroom fan, if you are sensitive to molds, it is a good idea to wipe down the tub and even the walls after showering.

Most of us spend a minimum of eight hours a day in the bedroom. Mattresses may be treated with flame retardants, and permanent-press sheets treated with formaldehyde and other chemicals that give off fumes. Check your bedroom closets. Do they have strong odors or fumes? Some laundry detergents and fabric softeners may have cloying perfumes that cling to clothes. Dry-cleaned clothes give off very toxic vapors and should be aired, preferably outside, before you hang them in your closet or wear them. If mildew or mold is accumulating in your closets, it can be transferred to your clothes, causing sensitivities when you wear them.

Even the attic needs to be ventilated so that it does not accumulate moisture. While you're there, check the insulation. Is it well sealed, or can moisture enter the house through the ventilation system? How about the heating ducts? All joints need to be taped and sealed to keep insulation from entering the ducts and filtering into the house.

The type of heating system that you choose is also important. There have been a number of studies in Canada showing that oil and gas furnaces will often "back-draft"— the fumes will actually be drawn into the home instead of being drawn out through the chimney. Both heating and air-conditioning systems need to have their filters replaced often and have regular cleaning. Furnace ducts can be vacuumed by professional cleaners to remove much of the accumulated dirt and dust. The average home has ineffective furnace filters that are only good for catching big chunks of dirt. Carbon and other high-efficiency filters are now available and can reduce the amount of particles in your indoor air.

Resources

The EPA has published several publications on the quality of indoor air:

•*The Inside Story: A Guide to Indoor Air Quality.* This 32-page booklet gives an overall view of some of the potential problems with indoor pollution. Focus is on the home, but there is a short section on the office. Pollutants discussed include pesticides, asbestos, formaldehyde, radon, cigarette smoke, stoves, and heaters. Also included is a list of local and state government agencies that can advise homeowners or prospective buyers.

•*Directory of State Indoor Air Contacts.* This 129-page publication gives a fairly comprehensive list of state agencies in each of the 50 states that may be able to help you with specific questions concerning indoor pollution. Included are things like building-complaints investigations, complaints about building materials like paint, insulation, and other matters like gasoline or gas fumes, pesticides, radon, and fiber analysis.

•*Current Federal Indoor Air Quality Activities.* This 43-page booklet lists the EPA's various areas of research and programs it conducts. It includes the name of the office responsible for a particular area and gives contact names and telephone numbers.

•*Congressional Directory: Environment.* This is a 600-page directory published by Environment Communications. It purports to list all of the congressional committees and subcommittees (as well as members, aides, and staff) dealing with environment issues. The publication also has a word index listing about 500 environmental problems.

The first three publications are available by writing or calling the EPA, Air Division, Office of Air and Radiation, 401 M Street, SW, Washington, DC 20460: (202) 475-8470. For the *Congressional Di-*

rectory, contact Environment Communications, 6410 Rockledge Drive, Suite 203, Bethesda, MD   20817; (301) 571-9791.

For information on radon, state officials will provide the names of reputable companies doing radon testing. The EPA also provides a list of companies and publishes two free booklets on radon testing—*A Citizen's Guide to Radon: What It Is and What to Do About It* and *Radon Reduction Methods: A Homeowner's Guide.* You can get this information by contacting your local branch of the EPA, listed in the telephone book under the United States Government section.

If you need information on formaldehyde products, contact the Consumer Product Safety Commission at (800) 638-2772.

Part One
The Air We Breathe, the Water We Drink

Chapter 1
The Greenhouse Effect

Not very long ago reports that our lifestyles and consumption patterns were turning the earth into a giant greenhouse would have been met by skepticism, if not outright cynicism. Of course, there are a few cynics who persist in labeling the ever-mounting evidence as science fiction—but then again, how long did the tobacco industry, and even members of the medical establishment, insist that smoking presented no proven danger to human health? As well illustrated in the smoking debacle, opponents to change often couch their arguments in an aura of scientific legitimacy, claiming that there is "no proof, no definite scientific proof" that smoking is linked to disease, or in the case of the greenhouse effect, that certain pollutants are turning the planet into a hothouse. Fortunately, the public has started to become wise to these tactics. Elementary logic shows us that the only proof that can be absolute in environmental issues, particularly one as potentially damaging as the greenhouse effect, is irreparable damage to the earth. Rational scientists are realizing that while the scientific method is still the preferred model of proof, protocol should not be inviolate when the future of the planet is at stake.

With the greenhouse effect, the nation is beginning to mobilize on an issue while it is still at the probability stage for the first time in environmental history. This is an encouraging move, which may be heralding a big shift in American consciousness toward action-oriented problem solving and away from avoidance and denial.

While scientists have been discussing the greenhouse effect for a number of years, debates picked up considerably during the summer of 1988 when 100 degree temperatures and withering drought seemed almost undeniable proof that global warming had indeed begun. One of the first announcements came from scientists testi-

3

fying before a Senate hearing in mid-June 1988, when they suggested that the greenhouse effect could be contributing to the droughts that parched farmlands and resulted in the designation of 40 percent of the counties nationwide as disaster areas.

The 1988 heat wave prompted James E. Hansen, director of the National Aeronautics and Space Administration's Institute for Space Studies, to speak out, even though he admitted that by doing so he was risking his professional reputation as a cautious scientist. Hansen's clear and unequivocal statement that the greenhouse effect was indeed upon us, coming from a scientist of his reputation, caused the nation to stop and listen. Hansen's message, a combination of hard, scientific analysis and bottom-line urgency, had a huge impact on political thinking. Bipartisan congressional support grew to allocate financing for climate research and to draft legislation for control of greenhouse-producing gases emitted into the atmosphere.

Commenting on the impact of Hansen's statements, Michael Oppenheimer, an atmospheric physicist for the Environmental Defense Fund, said that he's never seen an environmental issue take hold so quickly. "It took a government forum during a drought and a heat wave and one scientist with guts to say, 'Yes, it looks like it has begun and we've detected it.'"

Let's Talk About the Weather

In simple terms, the greenhouse effect can be defined as yet another way in which our planet is adversely reacting to air pollution. More specifically, scientists now believe that certain gases are concentrating and forming a barrier that admits the sunlight but traps in the sun's heat, which would otherwise be released into space, like the glass in a greenhouse. The overall effect is a global warming predicted to range between 3 and 8 degrees over the next 100 years. Temperatures have already risen more than 1 degree over the last century. This may not sound like very much of an increase, but on a global level small changes in average temperature can drastically alter the face of the planet. During the ice age when the planet was almost entirely covered with ice, the global temperature was only 8 degrees cooler than it is today.

To those of us living in colder climates or who simply "can't stand the cold," a global warming still may not sound like such bad news. But scientists warn that increases in temperature will be erratic and unevenly distributed around the globe. The greenhouse effect will not necessarily mean noticeably milder winters for quite some time, but it could, as was seen in 1988, make hot summers even hotter and dry seasons even drier. In addition, two researchers at the Massachusetts Institute of Technology note, "Greenhouse theory does not preclude extreme cold or extreme heat in any particular place or time," so that even unusual cold spells like the freezes in Florida are consistent with the theory.

Scientists also warn that while there may be benefits, the negative factors associated with such a global rise in temperatures will be far more predominant. In a draft report prepared by the normally cautious Environmental Protection Agency (EPA) in October 1988, the predictions of chaos and disaster were unprecedented. Global warming caused by industrial pollutants will diminish forests, destroy coastal wetlands, and cause extensive environmental damage over the next century.

The report concludes, "Global climate change will have significant implications for natural ecosystems; for when, where, and how we farm; for the availability of water to drink and water to run our factories, for how we live in our cities; for the wetlands that spawn our fish; for the beaches we use for recreation, and for all levels of government and industry."

Surprisingly, the primary culprit in setting off this thermal time bomb is not a highly noxious or toxic gas, but rather carbon dioxide, a naturally occurring component of the earth's atmosphere. Consequently, one lesson to be learned from the greenhouse effect pertains to the delicate balance of environment—even substances that are not normally pollutants can become harmful when their concentrations increase beyond levels that the earth and atmosphere can accommodate. Other gases suspected of combining with the carbon dioxide to create this hothouse effect are already notoriously linked with ecological damage. The following gases also have many times more capacity to trap heat than carbon dioxide:

•nitrogen oxides, one of the components of acid rain with 250 times the heat trapping capacity of carbon dioxide;

•chlorofluorocarbons, which are recognized as the primary offenders in the destruction of the earth's protective ozone layer, with up to 20,000 times the heat-trapping capacity; and

•methane, a gas released during the decomposition of organic matter with 25 times the heat-trapping effect per molecule.

The levels of these gases increase each year. Prior to the industrial revolution, the levels of carbon dioxide in the air were 275 parts per million (ppm). Today, the concentration has increased to 346 ppm, and is increasing annually at a rate of 0.4 percent. Similarly, methane's preindustrial level has more than doubled, nitrogen oxide's is increasing at a rate of 0.2 percent per annum, and fluorocarbons, which were not even known prior to the industrial revolution, are now increasing at an annual rate of 5 percent.

One of the major sources of carbon dioxide in the atmosphere is the burning of fossil fuels. Scientists are now saying that the impetus behind the global warming is so great that even drastic reductions in the burning of fossil fuels will not halt the trend from occurring. They also warn that without strong measures to decrease fossil-fuel combustion, changes could come upon us so suddenly and drastically that society might suffer a virtual breakdown. Climatologist for the National Center for Atmospheric Research in Boulder, Colorado, Stephen H. Schneider believes that the faster changes occur, the more "surprises and imbalances" we will experience. "The effects on forests, and of sea-level change, will be more damaging. But if the change is slow enough, you can learn how to adjust. You could develop seeds that will be able to take advantage of a longer but drier growing season."

Biding time will be particularly important for planning how to preserve our beaches, harbors, and coastal areas. Researchers studying the warming of the globe believe it likely that the melting of glaciers and the expansion of seawater as it is heated will cause ocean levels to rise, although nobody knows for sure how much . City officials in Charleston, South Carolina, in planning for a new storm sewer, are among the first to take these factors into account. EPA economists estimate that it could cost between $10 billion and $50 billion to salvage beaches washed away from rising tides. While most coastal cities could be protected by pumps and levees, hundreds of thousands of acres of lowland in Louisiana could end up underwater, and the port of New Orleans would have to be moved. Meanwhile, Dennis Tirpak, head of strategic studies for the EPA says that many roads, dams, water-supply systems, and storm drains are going to have to be designed keeping drastic changes in weather in mind.

The International Dilemma

Since the greenhouse effect is a global problem, one of the difficulties in reaching any sort of solution stems from the necessity of international cooperation. How could curbs in energy use be divided among those who use different amounts of the world's fuels? For instance, many of the world's developing nations, including China, base their plans for economic growth on the use of coal—the most plentiful of the fossil fuels, the cheapest, but also the one that releases the most carbon dioxide.

A related problem stems from the massive destruction of the tropical forests in the Amazon jungle. Thick clouds of smoke rise over the rain forest as the man-made fires of the annual dry season sweep the Amazon. On some days, thousands of fires roar across the Amazon basin, following a broad swath where settlers destroy the jungle frontier. For many, fire is the only tool for transforming forest into farm or pasture.

According to American and Brazilian scientists monitoring the Amazon, the fires are so numerous and the destruction so wide-spread that these annual burning rituals may be accounting for as much as *10 percent* of the global man-made output of carbon dioxide. Brazilian scientists estimate that in 1987 alone, 77,000 square miles of forest were burned, equaling an area about one-and-a-half times the size of New York State. Startling reports concerning the Amazon fires reveal that the resulting pollution has traveled thousands of miles. Among other things, these observations have led researchers to query whether the fires are related to the damage of the earth's protective ozone layer over the Arctic.

Brazilian researcher Alberto Setzer, coordinator of satellite data collected at the Space Research Center in São José Dos Campos, points out that this deforestation is doubly damaging—not only do the fires add tremendous amounts of carbon dioxide to the air, but by eliminating trees that utilize carbon dioxide as part of their photosynthetic process, the fires reduce the earth's capacity to absorb and neutralize the carbon dioxide that does exist in the atmosphere. The world is currently sending about 5.5 billion tons of carbon dioxide into the atmosphere, about half of which is absorbed by the oceans and forests.

According to a report issued by the Space Research Center, the fires in 1987 alone produced carbon dioxide containing more than 500 tons of carbon, 44 million tons of carbon-monoxide, more than 6 million tons of particles, almost 5 million tons of methane, 2.5 million tons of ozone and more than 1 million tons of nitrogen oxides and other substances that can circulate globally and influence radiation and climate.

The Brazilian situation provides a good example of the problems involved in coming up with solutions to the greenhouse effect on an international level. It is easy for us as Americans, living in an affluent society, to point our fingers and cry cut in dismay at what the Brazilians are doing not only to their own country, but also to the world. Not surprisingly Brazilian president José Sarney has denounced foreign meddling in his country's internal affairs and demands that developed nations clean up their own backyards before they concern themselves about their neighbors. Says President Sarney, "Brazil is being threatened in its sovereign right to use, exploit, and administer its territory. Every day brings new forms of intervention, with veiled or explicit threats aiming to force us to take decisions that are not in our interest." According to President Sarney, the industrialized nations are engaged in "an insidious, cruel and untruthful campaign" against Brazil, designed to detract attention from their own wanton pollution and "fantastic nuclear arsenals" that threaten all life on the planet.

Brazilian officials claim they are being blackmailed. They contend that foreign loans on the national debt are conditional, dependent on Brazil's taking adequate measures to protect the Amazon. While this may not seem unreasonable to us, the Brazilians are irate. One government official is quoted as saying, "There is true danger of foreign occupation of the Amazon. We are seeing a concerted international effort to hold back development in Brazil."

There is an irony in all this. One of the primary reasons for the felling of the Amazon in the first place stems not from Brazilian greed, but from the American appetite for beef. A World Bank study determined that official subsidies have made ranching in the Amazon artificially profitable. The study found that from 1975 to 1986 the Brazilian government in effect promoted deforestation with more that $1 billion in subsidies for ranchers. Subsidies were made to encourage export of Brazilian beef to the American market.

Save the World's Forests . . . and They May Save Us

Now, in addition to decreasing our consumption of red meat for health reasons, we have an ecological purpose too. We need to let the government and industry know that we do not want to participate in any way to encourage further destruction of the Amazon. It is not surprising that President Sarney was indignant about our demands to stop burning the forests when our own consumerism was the incentive for the burning in the first place. On the other hand, Brazil and other South American countries are developing nations, often strapped with enormous international debt. This requires them to produce goods for export. With a largely illiterate population, the most feasible products are agricultural goods. But agriculture requires land; hence, the burning of the forests.

Environmental issues that threaten our planet are no longer separable from economic development, consumption patterns, or any other aspect of life anywhere on the planet. If we are to solve or at least put a rein on the greenhouse effect, then the developmental needs of countries like Brazil need to be addressed in a responsible, not exploitative, manner. There is nothing wrong with conditioning debt concessions on adequate efforts to preserve the tropical forests, but such conditions cannot leave the debtor country without any means to support itself. For all too long we have treated the nations of South America and Africa as though they existed simply to provide us with an inexhaustible supply of natural resources. We are now seeing one of the results of this selfish and narrow-minded mentality in the destruction of the Brazilian forests.

How much do we value the forests of the Amazon? Do we cherish them enough to give Brazil and other Amazonian countries the money required to declare the forest a national- or world-heritage park? What if funds were then necessary for massive educational and training programs to enable today's peasants to support themselves without the land they are currently retrieving from the Amazon?

Ironically, the domesticated jungle does not provide very good farmland or even ranching land, which is one of the reasons that such massive amounts of forest continue to be cut—the peasants clear an are; then, when it no longer produces adequately, they move on to clear another area. So the question is, why aren't these peasants cultivating the good farmland? This, of course, raises all of the questions of politics in South America where a very few elite control 75 to 95 percent of a nation's assets. Often these governments receive political and financial support from the United States. The question of international politics and economics is obviously beyond the scope of this book. We merely mention this dimension to draw attention to the interdependence between our foreign policies and the environmental problems we are facing today.

Senator Robert Kasten of Wisconsin has said, "In my work on the Foreign Operations Subcommittee of the Senate Appropriations Committee, I have been shocked to learn the extent to which the United States has actually subsidized this kind of environmental devastation.

"If we're serious about preventing the global greenhouse effect from becoming a disastrous reality, we have to go to the source—and make sure our development lending stops financing ruinous projects. . . .

"The specter of the greenhouse effect will not be laid fully to rest until these concerns become an integral part of America's foreign assistance policy." /

A Simple Remedy

Destruction of the Amazon is having a doubly harmful effect—first, the huge fires release large amounts of carbon dioxide into the atmosphere, and second, the destruction of the trees decreases the earth's capacity to absorb and neutralize the carbon dioxide. In recognition of the natural capacity of trees to absorb carbon dioxide, scientists from the Environmental Defense Fund (EDF) have come up with a particularly brilliant—and painless—way to help slow down the global warming: Plant more trees. According to the EDF, an additional 10 million acres of forest would be able to absorb all the carbon dioxide emitted by power plants to be built in the next decade. Nor is the EDF proposing such a scheme without reasonable and feasible means of implementation. The environmental group suggests that the planting be undertaken by the Federal Conservation Reserve Program, which is a project of the Department of Agriculture. The Federal Conservation Reserve Program encourages farmers to lease land to the government for the planting of the trees. This accomplishes two additional benefits: the project targets lands, which are currently overproducing crops for which the government now pays subsidies, hence saving the taxpayer money, and also seeks out lands subject to erosion, which can benefit by the planting of the trees. The program is estimated to cost between $1 and $2 billion, much of which would be derived by increased assessments on the utilities building the new plants. Although this sounds like a large sum of money, it is by far the most cost-effective means to reduce the carbon dioxide in the atmosphere.

Another exciting pro-forest project is already under way due to a major funding grant from the MacArthur Foundation. The foundation gave more than $7.5 million to ten environmental groups working in Hawaii, Puerto Rico, the Virgin Islands, and the Florida Keys to acquire and preserve tropical forest areas in these regions. While these efforts are designed to protect the tropical forest in the United States and its protectorates, they also send positive messages to countries like Brazil that the Americans are at last willing to undertake responsible action themselves. Botanist Peter Raven, one of the leading advocates of rain-forest preservation, says, "The U.S. has got to express leadership in the preservation of its own tropical forests." He asks how we can expect other countries to preserve their forests while we continue to destroy our own.

Facing Up to Reality—Curbing Fossil Fuel Use
  We here in the United States have also actively contributed to the greenhouse effect by our paving of America. We have chopped down trees, blacktopped fields and meadows, and replaced this vegetation that thrives on carbon dioxide with asphalt roads for our automobiles. This trend must stop. Take the example of the Santa Clara Valley, about 40 miles south of San Francisco, where some of the world's best apricots, cherries, and prunes were once grown. Today, all those fruit trees are gone, the rich orchards paved over to make room for Silicon Valley. In typical California style, the automobile is the only means of transportation for the thousands of people who commute every day to and from various parts of the valley. As each acre of orchard was felled, new cars were added to the valley, until it became what it is today—asphalt-paved, grid-locked, smoggy, and a symbol of one of America's greatest contributions to the greenhouse effect.

  A number of years ago the rest of America could have watched that, as it watched Los Angeles, and said, "Well, that's their problem. If they want to live like that, let them go ahead." For most issues, that holds true. No one is going to help California out of gridlock until Californians are ready for it, but we cannot remain so blasé. The time has come for Americans to wake up to what has been abundantly clear for many years now—our dependence upon the automobile is destroying our environment and very likely our planet. The private car just isn't a viable means of transportation any longer.

  Much of the discussion about curbing the greenhouse effect has centered around the need to reduce carbon-dioxide emissions from utility plants burning fossil fuels—not decreasing dependence on the automobile. We need to realize that the vested interests of certain industries have worked long and hard to get us to our current dependency on the automobile. The destruction of the public-transportation system in Los Angeles is but one example. Likewise, the beleaguered nuclear industry, which has seen each of its plant orders canceled or abandoned during construction since 1974, welcomes predictions of global warming as the long-awaited answer to its business woes, evoking promises that only the nuclear industry can provide the panacea for global warming.

Very little is being said about what is possibly the single greatest contributing factor to the greenhouse effect. Transportation-systems consultant Jack Brannan is one of the few who speaks clearly on this point. He says this country's greatest source of air pollution is the internal-combustion engine, which powers all our private automobiles and intercity trucks and aircraft. "The number-one air polluter in North America is the private automobile, not the coal-fired power plant. . . .The automobile consumes almost forty-five percent of all the petroleum used in our country every year... The internal combustion engine's waste products are mainly oxide gases of carbon and nitrogen, plus water vapor. They contribute more than fifty percent to the carbon dioxide content of the air and the acid in acid rain!

"So, if anything is to be accomplished in controlling the greenhouse effect, plus acid rain, plus smog, petroleum use in transportation will have to be materially reduced.

"Proper employment of rail mass transit and rail freight can accomplish this reduction in the burning of fossil fuel, especially oil."

As in every debate about issues requiring change, discussions of a needed decrease in our consumption of fossil fuel draw varying reactions. Most of these reactions can be divided into two camps: the rational and the hysterical. These in turn seem to coincide with the different approaches to the problem.

The rational camp is largely composed of those people who have chosen to study all the issues carefully, rather than avoiding some or all of them or pretending that they do not exist. Their approach to the greenhouse effect is reasonable, measured, and practical. Their discussions go to the root of the problem and look at long-term, common-sense approaches to solving them. Irving Mintzer, a scientist for the World Resources Institute, a Washington, D.C., research center, believes that cuts in carbon-dioxide output can be made relatively painlessly and in ways that are beneficial to us in the long run. His proposals include automobiles that get 40 miles per gallon rather than the current 20, more efficient light bulbs and electrical appliances, urban planning that reduces traffic congestion, and a halt to deforestation in the tropics.

Equally constructive and realistic is atmospheric physicist Michael Oppenheimer of the Environmental Defense Fund. He endorses a solution proposed in a report by two United Nations agencies and a subsequent Canadian government conference: "Use fossil fuels efficiently in the short term, while investing in new energy sources for the long term."

According to Oppenheimer, efficiency is an important first step in moving to curb our use of fossil fuels. "The United States' economy grew thirty percent over the last fifteen years," he says, "but net energy use hardly budged. A United States carbon-dioxide-emissions reduction target of twenty percent in the next fifteen years could be reached by further increases in efficiency and some substitution of natural gas for coal and oil."

More is required than efficiency, however. "Only a gradual switch to renewable sources can fuel industrial growth while capping warming at tolerable levels," says Oppenheimer. Moving toward nonfossil fuel, renewable energy sources demands the expenditure of research and development dollars right now. Otherwise, says Oppenheimer, "[I]n a decade or two the world will be faced with a Hobson's choice: switch to a full-blown nuclear fission economy or suffer the warming. Given the problems dragging down the nuclear-power industry—economic, safety, waste, and weapons-proliferation issues—the development of renewable energy sources and the diffusion of technology to the fast-growing world is the wisest choice."

Oppenheimer believes that an efficient nonfossil-fuel, nonnuclear society is the unavoidable way of the future. He foresees the use of photovoltaic or solar power to generate electricity; and as to those automobiles that still run on gasoline, they will be averaging 100 miles per gallon. To the skeptics he replies, "A nation that could put a man on the moon or build the bomb in less than a decade can afford a Manhattan Project for alternative energy. America can either position its industries for the future, or end up buying energy technology elsewhere, as we have done with the high-milage auto."

As illustrated by Mintzer and Oppenheimer, people in the rational camp take a hands-on pragmatic approach to the problem. They come up with positive and constructive solutions grounded in good common sense. Persons in the opposing camp are noticeably more hysterical and reactionary. To them, doom and gloom is upon us, nothing can be done without totally uprooting society. In short, their message is one of disempowerment, one of being overwhelmed, and for the most part, at least a portion of their bottom line solution lies in nuclear power. There is a consistency in this. Those who are reacting hysterically now are those who have been avoiding and denying the existence of the problem. They are the ones who adhere to the need for more time to study the problem, for more concrete proof. They stall, and find obstacles instead of solutions. Now that they are finally grappling with the problem, they adopt the same irrational and short-term approach that caused them to avoid the problem in the first place.

The Rebuttal

The arguments of the nuclear industry are persuasive. One industry spokesperson says, "It seems that the major solution that's available presently is to stop burning. The only one you're really left with in a very pragmatic sense is nuclear." Other spokespersons tell us of a new generation of nuclear-power plant that has more backup safety systems and is "user friendly." Senator Timothy E. Wirth, a Colorado Democrat, who has introduced a bill to spend $4.3 billion to study the greenhouse effect, says that it is time for the country to get over its case of "nuclear measles." His bill provides $500 million for research on a safer nuclear reactor. "We have to rethink nuclear," he says. "We really have to start all over again, but we've got time to do it. The environmentalists will come around. They can't help but come around."

Readers need to remember what has been said previously:

•that power plants represent only a part of the problem, the use of the internal-combustion engine and the burning of the tropical forests playing very significant roles as well;

•that even when we are talking about finding solutions to the carbon-dioxide emissions of power plants, there *are* other alternatives to fossil and nuclear energies.

Photovoltaic energy is a viable alternative, and its costs are dropping. Given the real costs of nuclear energy—safety, waste disposal, and nuclear-weapons proliferation—is it really true that it is our cheapest and most viable alternative? Significant inroads have been made by engineers at the Electric Power Research Institute, a nonprofit group based in Palo Alto, California, in the fabrication of photovoltaic cells to suggest that the technology will soon be competitive with the more expensive fossil-fired electric plants.

Efficiency is also a very important factor. Says Scott Denman, director of the Safe Energy Communication Council, a coalition of environmental and antinuclear groups in Washington, D.C., "Today we spend ten percent of our GNP on energy. Japan, our primary competitor in many areas of the economy, spends only five percent on energy. If we want to mine for oil, Detroit is the biggest field we have."

There are many different scenarios about what life on a warmer planet would be like. Some scientists are saying that the greenhouse effect is upon us, others are making a last-ditch effort to rebut the evidence. In January 1989, after the flurry of news stories on the greenhouse effect throughout the summer and fall of 1988, *The New York Times* ran a story that claimed that "researchers agreed" that the greenhouse effect was not the cause of the 1988 drought and heat waves. "While most climate experts believe that the greenhouse effect . . . will have a major impact in the decades ahead, even those who argue most strongly for that point of view agree that last year's drought was overwhelmingly a product of natural forces. Even if a warming trend is already under way, as some experts believe, on a year-to-year basis events such as last year's changes in the Pacific exert a far stronger influence on the weather, scientists say."

It appears that scientists who set their mind to it can disprove anything. That is not to say that their theory is not correct. The theory about which "researchers agree" pertains to a strip of abnormally warm water that sometimes stretches westward along the Equator from South America. When this stretch of water, El Niño, comes into contact with another stretch of abnormally cold water, then all sorts of havoc break loose—trade winds can collide in unusual places, storms can result that disturb the atmosphere and create high- and low-pressure systems, which can in turn cause a lack of rain.

It is not clear why scientists can be so certain that all of these climatic *coincidences* are purely *natural* phenomena, unrelated to the greenhouse effect. Most scientists who have spoken on the effects of a warming planet indicate that all sorts of climatic, oceanic, and atmospheric changes can take place. While they have computer models of various scenarios, these models are ultimately man-made and cannot anticipate every variable.

Nineteen eighty-eight was a year of abnormally drastic weather. Apart from what we were experiencing in the United States, an unusually heavy monsoon season brought record floods to Bangladesh, resulting in death and misery. Australia also had unusual rainfall. The outback was greener than ranchers ever remembered seeing it, but the heavy rains damaged crops; and while rains in New South Wales did not reach flooding proportions in the spring and summer, 1988 was one of the wettest years on record.

We now have a choice. Either we can accept the claims of scientists that the greenhouse effect had nothing to do with this unusual and often ruinous weather, or we can address the inevitable. Even if the droughts, heat waves, and monsoons of 1988 were due to phenomena wholly unrelated to the greenhouse effects, all of the factors contributing toward a warmer planet are still in effect. Just because natural phenomena can evoke the same results does not mean that carbon dioxide and other gases in our air are not still building up and trapping the radiation from the sun close to the planet and the lower atmosphere. Scientists have already documented a global warming over the last century. As the planet heats up, scientists admit that they do not yet know how the oceans will react. NASA's James Hansen has assumed for the moment that nothing unusual will occur. "The ocean is an uncertainty," he says. "We don't have a good model for ocean circulation. But we assume the ocean will continue to operate in the near future as it has in the past." But Stephen Schneider, a climate specialist from the National Center for Atmospheric Research in Boulder, Colorado, believes that oceanic changes will turn out to play a very important role. According to another specialist, Wallace Broecker of Columbia University, if the seas do change dramatically, "all bets are off."

Even Hansen's model does not mean that we will see no changes. As the globe heats up, surface temperatures will rise in the oceans as well. That factor alone can cause the destructive force of hurricanes to increase. According to Kerry A. Emanuel of the Massachusetts Institute of Technology, the energy in a hurricane results primarily from two different factors: the difference in temperature between the surface waters of the ocean and the upper atmosphere, and the difference in pressure between the normal atmosphere and the air mass within the hurricane. Normal atmospheric pressure is about 1,013 millibars. The lowest pressure ever recorded within a hurricane is 870 millibars. Emanuel warns, however, that if the temperature of tropical waters rises, "the minimum pressure possible in hurricanes in the tropics may go down to as little as eight hundred millibars. That would mean the destructive potential of the most intense hurricanes would go up by forty or fifty percent."

We know that we are emitting more carbon dioxide into the atmosphere each year, and that we are destroying tropical rain forests, thereby lessening the earth's capacity to absorb that carbon dioxide. We also know that the earth's temperature has risen one degree in the last century, and that carbon dioxide combines with other gases to trap the sun's radiation and there by cause a global warming of the planet. According to Hansen's computer models that means that "the probability of having a hot summer will go up from thirty percent [as it was in the 1950s] to between sixty and seventy percent in the 1990s. That's enough so the man in the street will notice we're having a lot of hot summers."

Hansen believes 1988 was about one degree Celsius above normal for the continental United States. That should be typical about ten years from now. You can see that 1° is a significant warming.

So, the question remains: Is the greenhouse effect upon us, or isn't it? Or does it really matter? Our environment is giving us very strong messages that our cooperation is needed if it is going to continue to provide us with a safe, hospitable place to live. Many of the changes necessary to curb a trend toward a warmer planet are changes whose time have come anyway—decreasing our dependency on fossil fuels, on the automobile, preserving our rain forests, participating on an international level to find ways to growth that do not endanger our environment. Why not learn our lessons now and make life easier for ourselves, future generations, and the planet?

It is a mistaken belief that the changes are going to bring hardship and deprivation upon us. Thousands of people in this country, for instance, have already moved away from a meat-and-potatoes diet to one rich in grains, fruits, vegetables, and legumes. They are healthier, happier, and have no regrets about leaving red meat behind them. Many more people would readily abandon their car for mass transit to commute to work each day if such transportation was made available to them. Most would not even object to a reallocation of tax dollars or even a tax increase to subsidize transportation projects. Americans have had very little trouble adapting to smaller, more energy-efficient cars throughout the 1970s and 1980s. In fact, with many people finding life much more expensive these days than it was in the past, they would welcome any innovation that would save energy and money.

It is time for policymakers to realize that the American public is not a gigantic inertial mass that constantly resists change. As Michael Oppenheimer believes, if American policymakers and industries refuse to respond to current demand, we will end up having to go abroad to purchase new technology.

America has some of the most spectacular national parks, coastal areas, and wilderness areas in the world. We do more fishing, backpacking, cycling, jogging, and other outdoor activities than most other nations do. We lead the world in the pursuit of health, fitness, and exercise. Our environmental groups and conservation efforts are among the best anywhere. America sets trends around the world in music, medicine, film, sports, and environmental awareness. Isn't it time for us to show real leadership to the world when it comes to cleaning up our environment and developing the technology of the future—the technology that aims at preservation rather than destruction of our wonderful planet?

Hands-on Solutions

We interviewed Alden Bryant, president of the Earth Regeneration Society, who feels that our food supplies and farms are in serious trouble because of the heat, drought, freezing, and overall climate destabilization caused by the greenhouse effect. About 170 billion tons of CO_2 need to be taken out of the atmosphere within 15 years in order for us to get back to our regular living conditions and to prevent our foods from being destroyed.

Bryant suggests we support the bill presently in Congress to establish a carbon-dioxide budget. It will monitor the amount of CO_2 produced in your area and promote action to help start stabilizing the climate. The California CO_2 budget is presently a model for the United States and other countries in how to lay out a local jobs program around environmental problems resulting from the greenhouse effect.

We may forget that photosynthesis provides an easy remedy, at least in part, for the greenhouse effect dilemma. To make up for the trees that have been felled in the rain forests of South America and elsewhere, why not plant some additional greenery here at home? In the process of growing and producing their own food, plants and trees absorb CO_2 from the atmosphere and "breathe out" oxygen. By planting some trees in your yard, near the street, in a park, or even some flower pots on your windowsills, you can help the effort while adding to the natural beauty that surrounds your home. City and state nurseries can provide seedlings free of charge.

Farmers should be encouraged and helped to plant more trees around farm areas. In addition to stabilizing the oxygen-carbon dioxide balance, they can also protect the land from wind erosion. Farmers should use good rock dust to remineralize soil and bring it back up to good natural conditions; to save trees that are dying, add 20 to 50 pounds of good gravel dust or rock dust to the surrounding soil.

The best solar electric utility in the world is in southern California and supplies power to between 270,000 and 280,000 people. It costs under eight cents a kilowatt-hour and is supplied by the Luz Corporation. This kind of technology can serve most of the country's electrical-power needs. Use of solar-powered technology can also help cut down on fossil-fuel use and acid rain.

More information can be obtained from:

Earth Regeneration Society
(415) 525-4877

Chapter 2
Ozone Depletion

All environmental problems we face today are somehow related. There is a general lesson to be learned from the state of our planet as a whole. It goes something like this: for many years now we have treated our planet as an inexhaustible resource. We have acted as if there were unlimited air, water, oil, and forests. Now, the planet is letting us know that it cannot continue to be treated in this way.

Each environmental issue, while being part of this larger scheme, also has its own message to teach us. In the case of the depletion of the ozone layer of our atmosphere, we are learning that even the most innocuous of chemicals, chemicals that are almost completely nontoxic to humans and cause no damage to the earth per se, can wreak havoc miles above us in ways we could never have imagined. On a more encouraging note, there is something else that we may learn from this particular problem. Contrary to what industry spokespersons and some government officials are saying, even a total elimination of these chemicals will not necessarily be painful or require great sacrifice. There are already viable alternatives to ozone-depleting chemicals that are safe, inexpensive, and just as effective. While we may not be able to undo the damage that has already been done, moving to prevent further damage may turn out to be much easier than we have been led to believe.

The Earth's Protective Shield

When we talk about "ozone depletion," the type of ozone to which we are referring is found in the upper reaches of the atmosphere, some 10 to 30 miles above the surface of the earth. Atmospheric ozone is a gaseous compound made up of three oxygen atoms. It is formed in the stratosphere as a result of the interaction between the sun's ultraviolet rays and regular oxygen molecules, which contain two oxygen atoms. The ultraviolet radiation causes the oxygen molecule to split into highly reactive individual atoms that then recombine, some in pairs to form oxygen, some in trios to make ozone.

The primary importance of ozone in the upper reaches of the atmosphere is to shield the earth from the potentially damaging effects of the sun's ultraviolet rays. In the absence of this shield, life on Earth would be fairly close to impossible. Brian Toon, an associate fellow at the National Oceanic and Atmospheric Administration, says, "Without ozone, we'd all have to go back to living under the water."

According to William Walsh, staff attorney for the U.S. Public Interest Research Group, a nonprofit environmental organization, the depletion of the ozone layer currently taking place is, with the possible exception of nuclear weapons, one of the gravest crises we face here on Earth. Unlike nuclear war, atmospheric devastation is occurring right now. Despite extensive tests and computer modeling, scientists are still unable to determine whether it is too late to preserve the atmosphere as we currently know it for a sustainable and livable world in the future.

We can begin to understand the importance of the ozone layer by looking at just how essential it is to life as we know it. According to Walsh, "If we saw a decrease of the ozone layer of seventy percent, for example, for a Caucasian to go out into the sun would result in blistering sunburn after only ten minutes of exposure. If there was a fifty percent decrease, a Caucasian would have about an hour in temperate climates before he or she would experience a sunburn to the point of blistering. Even with only a twenty-five to thirty percent decrease in the ozone, outdoor work such as farming, fishing, and building would be virtually impossible because of the risk of sunburn and resultant skin cancer."

If we continue to destroy our ozone layer as we have done over the past 30 to 40 years, there will be a gross increase in the number of skin cancers and other cancers, likely food shortages, ecological disasters, and a variety of other health injuries we have never had to deal with before because we have never been exposed to such great amounts of ultraviolet radiation. There would also be increased weakening of our immune systems. Increased exposure to ultraviolet radiation would decrease our body's ability to stave off other illnesses, an effect similar to the AIDS virus.

Moreover, a reduction in the ozone layer could greatly exacerbate global warming already under way due to the greenhouse effect. A radical loss of ozone might possibly render the world uninhabitable. The stratosphere could heat up by as much as 40 degrees Fahrenheit, warming air close to the earth by about 4 degrees and melting the polar ice caps. Sea levels would rise and submerge many coastal cities. Rainfall patterns could change radically, flooding the deserts, and turning the north-central United States into a dust bowl. A recent study by NASA scientists estimates that 20 percent of the greenhouse effect can be attributed to the destruction of the ozone layer caused by man-made chemicals. Furthermore, some types of chlorofluorocarbons (CFCs) have up to 10,000 times the heat-holding capacity of carbon dioxide and hence are among the most potent of the greenhouse gases.

Even with relatively minor decreases in the ozone layer's protective capacity, scientists predict that there would be sharp increases in skin cancer, eye damage, and immune-system suppression in humans, while the increase in ultraviolet radiation could also endanger forests, crops, and wildlife. Experts at the EPA estimate that for every 1 percent decrease in ozone, skin cancer could increase by 5 *to 6 percent*.

History and Theory Behind the Depletion

The production and destruction of ozone in the upper layers of the atmosphere is a continuing natural process that results from the interaction between oxygen and ultraviolet light in the stratosphere. During the last 50 years or so, with the introduction of CFCs, what we are now seeing is both an unnatural and escalated destruction of the ozone layer that protects the earth.

In the mid-1970s two American scientists began to research the group of relatively inert chemicals called chlorofluorocarbons. Their interest in these chemicals was sparked by the findings of a British atmospheric chemist, James E. Lovelock, which revealed traces of the chemicals in analyses of the atmospheric samples taken over the Irish Sea. Lovelock was not particularly concerned about his findings, as the CFCs were reputed to be extremely stable and nontoxic. For this reason, they replaced a number of other chemicals for use in a wide variety of products—as coolants in air conditioners and refrigerators, as propellants in aerosol sprays, as foaming agents for things like automobile seats and dashboards, in certain types of plastic materials such as Styrofoam, and more recently as cleaning agents, particularly for computer circuitry. It was precisely this chemical stability that bothered the two American researchers. If the CFCs were not broken down and remained intact in the troposphere, the layer of the atmosphere closest to the earth, then what happened to them? Did they just float around in deep space for eternity?

The scientists, Mario J. Molina, now of the Jet Propulsion Laboratory in Pasadena, California, and F. Sherwood Rowland of the University of California at Irvine, discovered that the CFCs were not innocuous at all. CFCs were so stable that they floated unimpeded through the troposphere and into the ozone layer.

Molina and Rowland discovered that while CFCs do not degrade or chemically alter in the lower atmosphere, upon entering the stratosphere, where they are exposed to the ultraviolet radiation of the sun, they break down to their component atoms of chlorine, fluorine, and carbon. Although scientists are still not exactly sure what occurs after that, most believe that it is the chlorine portion of CFCs that goes on to damage the ozone layer. Furthermore, these chlorine molecules in the upper atmosphere can react many times over, and consequently a single molecule of chlorine can destroy thousands of molecules of ozone.

Around 1982 the CFC hypothesis moved from the realm of theory to that of substantiated scientific fact. At that time the British Antarctic Survey at Halley Bay, a group of scientists who had been monitoring the atmosphere in the Antarctic since 1957 suddenly discovered a large hole in the earth's ozone layer over the South Pole. Since then scientists have noted that every year around springtime (sometime between mid-August and mid-October in the Antarctic), huge gaps in the ozone occur. In September 1987, the ozone shield had dwindled to its lowest level ever observed at that time. Scientists measuring the ozone levels found that at an altitude of 11 miles the ozone shield had been reduced by 50 percent. Later that same year National Science Foundation official Peter E. Wilkniss reported to Congress that scientists at the McMurdo Station in Antarctica had observed that ozone levels at some altitudes had dropped as much as 97 percent. The serious depletion caused Wilkniss to express concern "about the health and safety of our workers in Antarctica, who may be exposed to as much as four times the amount of ultraviolet radiation as you would get in summer at the beach in Miami."

Walsh describes this hole as a gaping tear across the upper atmosphere that is as large as the continental United States and as deep as Mount Everest is tall. Additionally, the scientists found high levels of active chlorine in their samples of the atmosphere.

Findings above the McMurdo Station indicated levels of chlorine monoxide to be up to 100 times greater than those found elsewhere.

Robert Watson, chief scientist for NASA's ozone project, believes that there is no question that there is chlorine in the Antarctic in quantities enough to destroy a dramatic level of protective ozone.

When scientists first discovered the hole in the ozone, they were cautious not to extrapolate their findings to other parts of the globe. Watson, for example, posited that the ozone depletion was probably due to the peculiar atmospheric conditions at the South Pole; he was still uncertain whether there would be a worldwide fallout. Most scientists admit that they just do not yet know enough about what goes on in the high reaches of the atmosphere to be able to say with any certainty what is going on. Theoretically the depletion that is seen in Antarctica could have occurred anywhere. One of the great problems with ozone depletion is that we do not know enough about upper-atmospheric science or the chemical reactions between ozone and chlorine to really predict with any degree of accuracy what we can expect from our continued loading of chlorine into the atmosphere. The same NASA report that noted the gaping hole in the ozone above Antarctica also concluded that those recorded losses were more than twice the amounts that had been predicted by computer modeling in the years prior to the satellite survey. No one can explain these great losses. Today's depletion ratio, both in Antarctica and above the continental United States, was originally the rate of depletion that was predicted for the next century. So we are on a very fast track—much faster than anyone had dared to project when ozone depletion was first discovered.

Scientists now believe that CFCs are instrumental in the depletion of ozone at all latitudes, but weather conditions at the poles do contribute to the problem. Atmospheric measurement over Switzerland, South Dakota, and Maine all show some ozone depletion in late winter or early spring. This is confirmed by findings showing ozone losses since 1969 as high as 3 percent over parts of North America and Europe and as high as 5 percent over parts of the southern hemisphere. This is particularly disturbing given the EPA prediction that each percentage decrease in ozone could increase incidence of skin cancer by 5 to 6 percent.

We also know now that massive ozone depletion is not confined to the South Pole. In October 1988, scientists in the Arctic made findings similar to those of their southern counterparts. Evidence points to the poles as being particularly susceptible to a sort of accelerated ozone depletion, especially when compared to the changes which are taking place much more slowly elsewhere on the planet.

The process begins when the CFCs are broken down by ultraviolet rays primarily over the tropics and temperate zones. Global winds then tend to blow these component parts toward the North and South Poles, where the extreme climatic conditions come into effect. The CFC by-products adhere to airborne ice crystals during the winter. In the spring, when the crystals are hit by the sun's rays, the ultraviolet light causes the chemical reaction between the chlorine and the ozone. While the ozone is destroyed by the reaction, the pollutants survive to cause more destruction, over and over again.

Corporate Stalling— Obstacles and Solutions

When chlorofluorocarbons were first introduced in the 1930s, they were considered chemical marvels. Because of their stability, CFCs were almost completely nontoxic to humans, as well as being nonflammable and noncorrosive. In fact, they were considered to be so safe that, up until the findings of Rowland and Molina, CFCs were called upon to replace a number of more hazardous chemicals.

But that was before. What happened after CFCs became linked with damage to the ozone layer provides a good example of how corporate America can react when a source of its revenues is threatened.

Although the United States accounts for only 5 percent of the world population, it is a major contributor to the use and production of ozone-depleting CFCs. We produce close to 30 percent of the 2.4 billion tons of CFCs manufactured in the world each year. The Du Pont corporation, the largest worldwide manufacturer of these chemicals, controls approximately 45 percent of the U.S. market and about 25 percent of the world market. In 1974 there was a great debate as to whether CFCs should be banned from aerosol cans. Du Pont took out full-page ads in *The New York Times* and *The Washington Post* urging that there was not enough evidence to warrant such drastic action as prohibiting the use of the chemical in the aerosols.

At that time the company had begun research to develop substitute chemicals that could be used as replacements. In 1978, when aerosols were banned, Du Pont and other corporations cut back on their research, virtually eliminating development of alternative chemicals. They stated that the regulatory climate in the United States indicated there would be no government mandate for a future ban on the chemicals and therefore they did not intend to pursue their research into alternatives.

Now, we've lost about five years of research into alternative ways of cooling air, keeping refrigerators running, and cleaning high-tech electronic equipment. The rush to find alternatives to replace CFCs now could result in the use of other toxic chemicals.

Despite all the posturing and resistance, industry does come around when it has to. But is this all really necessary? Where is the government leadership that dictates policy *to* the chemical industry? By 1988, E. I. Du Pont de Nemours & Company announced that it would phase out production of CFCs by the year 2000. The Institute for Energy and Environmental Research, a Maryland-based public-interest group, among others, believes this action isn't enough. Chemical companies have taken the reins from government. It's time that the people force the policymakers to take back control.

Policymakers have made some inroads to curbing the use and production of ozone-harming chemicals, but while they use strong language about the need for a total ban, very little is being done to implement that sort of program. According to an EPA study released in September 1988, an immediate 100 percent reduction is required just to stabilize the chemicals already in the atmosphere. Even if the terms of the Montreal protocol, which call for a 50 percent decrease in production by the year 2000, are met, NASA's Robert Watson believes that the level of CFCs in our skies will double over the next 50 years.

Painless Solutions

Reactions to the depletion of the earth's protective ozone layer have been amazingly swift, given the usual inertia of government and industry to environmental issues, and shockingly slow in light of the severity of the problem.

First, a positive note. The American government was the first in the world to place an outright ban on the use of CFCs in aerosols in 1978. That was done only a few years after the work of Rowland and Molina on ozone destruction by the chlorofluorocarbons, and *before* the huge hole in the ozone layer was discovered in Antarctica in 1982. Despite the predictions and warnings from CFC manufacturers that Americans just could not survive without their aerosols, most of us weathered the transition without the least withdrawal effect.

One important point that is often overlooked in the debate about abolishing, or at least significantly decreasing, production and use of CFCs is that their importance to society may be more hype than reality. Unlike oil, which powers our automobiles and power plants, and for the moment really is a necessity in society pending the development of alternatives, chlorofluorocarbons on the whole serve convenience functions only. The choices we face when looking at the end to CFCs are no more profound than choices like these:

• aerosol versus roll-on deodorant;
• Styrofoam versus recyclable paper containers for our fast food;
• CFC as opposed to soap or other nontoxic cleaning products for computers.

Even in situations that may represent *necessity* rather than convenience, like refrigeration, air-conditioning in certain climates, and drycleaning of certain fabrics, carefully designed machines that lock in CFCs could substantially decrease, if not eliminate altogether, the emission of the chemicals into the environment. The one thing delaying a total ban is the insistence of the manufacturers that the world needs CFCs, or at least needs what CFCs can do for us. How true is that? America survived before the invention of aerosols, and is surviving very well now without them. Aerosol use accounts for 40 percent of European use of CFCs. Since Americans have decreased their use to less than 1 percent of their total CFC consumption, is there any reason to believe that Europeans will have any problems in doing so? How dependent are we on Styrofoam and other blown plastics? Are these products really indispensable, or is that just the manufacturers wishful thinking? In Asia and the Pacific Rim, 35 percent of CFC consumption is for cleaning purposes, especially the cleaning of electronic circuitry. As there are now safe and nontoxic cleaning agents for electronics, how indispensable are the CFC-based solvents?

Notwithstanding the United States ban on aerosols, other nations have yet to follow the American example. This leads to some serious questions of corporate responsibility. DuPont, for instance, is fully aware of the current scientific evidence concerning the destruction of the ozone layer by CFCs. Presumably it also knows that the problem is a global one. Yet the company continues to market its products to foreign countries for uses that are outlawed in this country. There is no moral justification for this practice on any level, but for even the most selfish and shortsighted Americans, the proof is now in that this affects us here at home every bit as much as it affects the Europeans or South Americans who are using the CFC-propelled aerosols. Unless we are looking for illusory solutions rather than real ones, a first step in stopping the use of CFCs worldwide would be to put a ban on production by U.S. companies.

In recognition of the need for a global approach to solving the ozone-depletion problem, representatives of 46 nations met in Montreal in September 1987 to discuss an agreement freezing the production of CFCs at 1986 levels and phasing in a 50 percent reduction in usage starting in 1989. To become effective, the agreement required ratification by eleven nations representing 66 percent of global consumption of CFCs. After a slow start, 31 countries had ratified the treaty by December 1988.

The United States has been more of a leader regarding ozone-depleting chemicals than it has on other environmental issues. In September 1988, prior to ratification of the treaty by the requisite number of nations, then head of the EPA Lee M. Thomas called for the complete elimination of CFCs and halons (another chlorine-based chemical linked to ozone depletion and used primarily in fire extinguishers). While the United States is talking tough, when it comes to action, Americans are still dragging their feet. Legislators have yet to act on the EPA chief's proposals. In the meantime, the twelve European Community nations, not the United States, who were the first to agree in March 1989 to a total ban on the production and use of ozone-destroying chemicals by the end of the century. William K. Reilly, head of the EPA, urged President Bush to endorse a worldwide ban, but for the moment the U.S. is vacillating—torn by environmental concerns on one hand, and by threats by industry representatives that substitute chemicals will be more expensive and not necessarily safe.

Another piece of good news is that contrary to what we are being told by industry and sometimes the press, there are viable solutions on the horizon that are neither costly nor energy intensive, and many of them are within the reach of the individual. Many alternatives to CFCs are already available, but what is lacking is a corporate commitment to make the switch and a legal mandate that would ban the production and use of atmospherically damaging chemicals.

The demand for CFC-based cleaning solvents in the computer industry has practically tripled since 1976—just about the time that CFCs were being banned for aerosol use. The United States produces about 40 percent of the worldwide supply of these solvents.

And while industry maintains that there are no viable substitutes for the ubiquitous CF-113, the most commonly used of these solvents, there has actually been great progress made in recent years toward very safe nontoxic solutions for cleaning electronics equipment that could greatly reduce the amount of CFC emissions. According to Walsh, one IBM manufacturing plant in California alone emits more than a million pounds of CFCs into the atmosphere annually. In recent years, high-tech companies like IBM, Digital Corporation, and AT&T have been experimenting with new aqueous cleaning solutions and have found them effective in cleaning electronic circuitry. They have also tried other newly developed products called terpenes, which are nontoxic, biodegradable, and noncorrosive substances derived from citrus rinds and pine trees. There are also techniques being developed that would allow for the design of no-clean circuitry that would not even require the use of solvents for maintenance and cleaning at all.

Since the ban on CFCs for aerosols, the single largest use for these chemicals is refrigeration, which accounts for 45 percent of all U.S. consumption of CFCs today. Both refrigerator manufacturers and the chemical industry are making claims about the problems associated with a move away from CFC-based refrigerants. In fact, if the rigid-foam insulation in refrigerators and freezers were replaced by vacuum insulation, the same type used in Thermos bottles, CFC use would plummet. The Solar Energy Research Institute has found that vacuum panels take up less space than foams and actually add to the efficiency of appliances.

Ideas for Personal Action

There are a number of other possibilities for refrigeration that are safe and just as efficient as CFCs, like helium-cooled refrigerators. Helium has long been used as a coolant in space and military application. One New Jersey-based company is currently adapting that technology for use in refrigerators. Redesign that would prevent all leakage of CFCs and legislation that would mandate the recycling of refrigerants when units are discarded would also eliminate the release of the CFCs into the atmosphere.

There are also a number of actions that each of us can take as individuals and as a group. The first and probably the easiest is to stop buying Styrofoam and similar foam products. These products are poison to the ozone layer and poison when they are burned. One of the largest users of Styrofoam are fast-food and take-out restaurants. Whenever you patronize these restaurants, you can begin your environmental action by requesting the operators to package your food or serve your coffee in containers that are made of something other than blown foam. The companies that make Styrofoam are moving toward other alternatives that do not require ozone-depleting chemicals to manufacture the substance, but it is impossible for the consumer to tell which has been manufactured safely and which has not. It is better just to stay away from the product altogether.

Something else the individual can do is to give some serious thought to the use of air conditioners, particularly automobile units, which virtually all use ozone-destroying chemicals. Many of us switch on the air conditioner in the summer merely out of habit; we would not even consider not buying an air-conditioned automobile. For those of you who live in relatively temperate climates, you may want to reconsider things like these.

In the United States, car air conditioners account for about 20 percent of the consumption of CFCs, the largest single share of gases contributing to the ozone-depletion problem. In fact, engineers estimate that about 65 percent of a car's air-conditioning fluids leaks before the car ever gets to the repair shop, and this leakage accounts for about a third of the emissions from automobiles. If you decide to buy a car air conditioner, never try to repair or refill it yourself. You should have it checked periodically and serviced by reputable professionals who use recycling equipment that enables them to get all the CFCs out of the air conditioner, fix it, tune it up, and then return CFCs to the unit without releasing any of them into the air. When looking for repairmen, you should ask whether they have such recycling equipment. The number-one action that you as a consumer can take is to make sure that your car air-conditioning system is not leaking any ozone-depleting gases. Second, when taking vehicles in for service, if you're told that the coolant from your air-conditioning system needs to be replaced, ask the service station about what will happen with the coolant that is drained from the automobile. The coolant is traditionally drained and vented into the atmosphere, leading to destruction of the ozone layer. Ask your auto mechanic about it.

While automobile and chemical manufacturers are gearing up for the testing of new air-conditioning chemicals, the news is that these chemicals are certainly not any great step forward. General Motors officials are saying that the new chemicals will require a total redesign of existing units, but they fear that fuel economy and design may have to be compromised to accommodate the change.

People should keep informed through their newspaper or call their local legislators, because approximately 20 states will soon be voting on legislation that would require recycling of car air-conditioner chemicals. We also encourage people to write or call not only their local representatives, but also their congressmen and senators, urging them to pass legislation to ban ozone-depleting chemicals rapidly and guarantee safe substitutes so that we do not find ourselves in yet another toxic crisis down the road a bit.

According to Cynthia Pollack Shea, senior researcher at World Watch Institute in Washington, D.C., we should avoid buying foods like eggs and meat packed in polystyrene foams. Ozone-depleting gases are used to make those bubbles. Ask the manager at your local grocery store whether or not those foam products are blown with ozone-depleting CFCs, and if they are, ask if trays can be purchased from another supplier. Request paper or cardboard wrapping.

When discarding old appliances that use CFCs as a coolant, find out what will happen to the old coolant. Will it be drained or recycled? What will happen to the old appliance? Will the insulation foam be covered and incinerated to keep the CFCs from reaching the stratosphere?

When you buy a home computer, ask if the computer chips need to be cleaned with a CFC-based solvent. Many currently are, but alternatives are available. You can clean computer chips with a water or alcohol solution, and some don't require cleaning at all.

When you buy furniture, ask whether or not the cushions in that furniture have been inflated with CFCs. Flexible foams are one of the major uses of fluorocarbons and alternatives are available for blowing the foam.

CFCs are also widely used in rigid-foam insulation. When engaging a contractor to install insulation in your home, ask whether or not that particular insulation is blown with CFCs. Use an alternative, fiberglass, cellulose, or foam that is not blown with CFCs.

The more questions you as a consumer ask dealers and repairmen, the more they will be aware of the high level of concern and their role in the problem.

For more information contact:

World Watch Institute
1776 Avenue N.W.
Washington, DC 20036
(202) 452-1999

Chapter 3
Acid Rain

The phenomenon known as *acid rain* has become one of the environment's worst blights. For almost 40 years scientists have issued reports on its devastating effects, and the public has watched as this by-product of the industrial age has destroyed forests and lakes throughout the world. Despite the controversy, very little has been done to implement solutions to the problem. In this regard, acid rain is no different than the multitude of other environmental problems we now face: The evidence is in, the public demands action, but the political factions continue to engage in specious debates designed to deny the all too obvious effects of the problem and to thwart initiatives that offer viable solutions.

<u>What Is Acid Rain?</u>

Acid rain results from certain kinds of air pollution that mix with precipitation, such as rain or fog, and then fall to Earth as an acidic solution. Its major components are oxides of sulfur and nitrogen that are primarily the by-products of coal-burning power plants, copper smelting, and factory and automobile emissions. These oxides are chemically altered in the atmosphere and return to the earth as rain, snow, fog, or dust. In the United States, the most popularly recognized form of acid rain results from sulfur-dioxide emissions, which are converted into sulfuric acid in the atmosphere. When this is mixed with precipitation and falls to Earth, the effect is precisely like pouring a diluted acid solution onto everything it touches. Although the effects of acid rain are most drastically felt and seen in the delicate ecostructures of our natural surroundings—the lakes, forests, farmlands, and oceans—its corrosive properties are now recognized to cause significant damage to even the man-made parts of our environment such as monuments and buildings.

In lakes, this acidification process can change ecological structures full of freshwater life into sterile, eutrophied dead zones. Particularly susceptible are those lakes with a low alkaline content or those that lack alkaline rocks and vegetation in their surroundings that can neutralize the acidic rainwater. Many of the nation's best trout-fishing spots have simply disappeared as a result of excess acidification. Once a lake has become too acid, bacteria and plankton die off, and little by little the entire food chain of the lake dwindles. Eventually frogs, fish, and other insects perish as well, leaving the lake itself barren and dead.

The Canadians, whose lakes in Quebec and Ontario have been particularly hard hit by acid rain, have been conducting unusual scientific experiments that enable us to see for the first time how acidification really works. By purposely releasing an acidic solution into a number of closely observed lakes in western Ontario, scientists are able to document the process on freshwater ecosystems. These studies are significant because they not only provide us with a time frame for the acid-rain damage, but also furnish the irrefutable correlation between an increase in the acidity level and the destruction of lakes and streams.

Some of the scientists' important findings include:

• The destruction of a lake's ecostructure can take place early on in the acidification process.

• The disappearance of adult fish is not an accurate measure of damage. Prey fish and reproductive dysfunction are the first indications that acidification is upsetting the lake's ecosystem.

• Lakes can begin to regenerate soon after acid input ceases, but it may be many years, perhaps even hundreds of years, before the original structure of the lake is restored.

• Wetlands can serve as buffers for lakes by filtering out acid-causing substances and thus can help protect lakes from acidification.

The National Audubon Society has monitored the pH levels in 39 states since July 1987. In January 1988, Audubon's Citizens' Acid Rain Monitoring Network issued its monthly report, which showed that almost half of the states experienced an average pH of 5.0 or less (on a scale of 0 to 14, with 0 being most acidic), even during winter when the water is generally at its least acidic. According to Audubon's senior staff scientist, Jan Beyea, this level of acidification is solely caused by air pollution stemming from man-made sources. Of all the states surveyed, Tennessee had the worst pH average with 4.2, while nine states (including Connecticut,

Maine, Massachusetts, Michigan, New York, Pennsylvania, Tennessee, Vermont, and Virginia) showed readings of 4.0 or less.

How Much Damage Is Acid Rain Really Causing?

For a number of years, much of the debate around acid rain in the United States centered on the freshwater lakes and streams in the Northeast that were quietly being destroyed in areas like New York's Adirondack Mountains and the Appalachian plateau of West Virginia. Much of the blame was laid on the sulfur dioxide from the industrial centers of the Midwest, which traveled often hundreds of miles before returning to Earth as acidic precipitation. We now know that acid rain is not merely confined to sulfur-dioxide emissions from a limited geographic area that affects another limited geographic area. It is an international problem that honors no boundaries, that results from a wide variety of air pollutants, and that has repercussions on our environment that we are only beginning to understand.

Acid rain has been around for just about as long as we have been polluting the air. In the 1800s, when large amounts of coal were being used, scientists in England first began to make the connection between sulfur oxide emissions and damage to surrounding plant life. In this century, it was not until the 1950s, in Norway and Sweden, that scientists began to publicize the correlation between acid rain and environmental damage. Today, acid rain is known to be prevalent in much of the industrialized world. In Europe, Japan, Canada, and the United States whole geographic areas of once-fertile natural resources are being doomed to a slow death by the sulfur dioxides and nitrogen oxides spewing forth from society's energy plants and factories. This damage is resulting in a great diminution of the world's heritage, for the destruction of the Swiss Alps or the corrosion of France's Gothic cathedrals deprives not only the countries in which these natural and historical treasures are located, but impoverishes us all.

Thirty years after the Scandinavians began documenting acid rain, the subject made its first major headlines in the United States when *Time* magazine reported in 1980 the disappearance of fish in about a hundred lakes in the Adirondack wilderness and labeled acid rain a "newly recognized and increasingly harmful kind of pollution, invisible and insidious." By 1982, there was sufficient uproar about the devastation caused by acid rain to warrant congressional discussion of amendments to the Clean Air Act and a study by the Office of Technology Assessment (OTA). The OTA report revealed that in the 27-state region covered by the study, 1 of every 4 lakes and streams in the northeastern United States has been damaged by acid rain. More specifically, the report stated that 1 out of every 6 lakes and 1 in every 5 streams had been damaged by acid rain, and that of 17,000 lakes in the area, more than 9,000 are endangered. In short, the report concluded that in the Northeast and upper Midwest, up to 80 percent of the lakes and streams are threatened with extinction by acid rain.

Notwithstanding the frightening message of the OTA report, little has been done to stop the proliferation of damage due to acid rain. In fact, it appears that, if anything, the situation is worsening. A recent Environmental Protection Agency (EPA) report shows that acidification of freshwater is much more widespread than was previously known.

The EPA previously found serious acidification in *only* several hundred lakes in New York State's Adirondacks and in parts of New England. An EPA survey of streams in the eastern United States revealed that about 4.4 percent of the 66,000 miles of stream in the middle Atlantic coastal plain stretching from New Jersey to North Carolina were acidic, and roughly half of the streams in the area were found to have a potentially low capacity to neutralize acid rain. This indicates a statistically proven basis for demonstrating a broader geographical extent of environmental effects from acid rain than was heretofore acknowledged. It could be just the beginning.

Scientists in California are finding that pollution from automobile exhaust and industries is endangering the lakes in the Sierra Nevada mountain range. A team of scientists recently issued warnings that Sequoia National Park's Emerald Lake had almost no capacity to neutralize future increases in acidity. In the past the lakes in the High Sierras have been relatively safe because, while rainwater has been about 10 percent more acidic than normal, snow, which constitutes more than 90 percent of the Sierran precipitation, was within the normal range. Now, scientists are discovering that the spring melt can sent "acid pulses" into the lakes, as most of the toxic chemicals are contained in the first 20 percent of the melt water. These acid pulses send the lakes into temporarily acid conditions, which most are now able to neutralize. Once a lake's natural neutralizing compounds are exhausted, as is the case with Emerald Lake, the acid levels can rise sharply and damage will ensue.

Nor is it just our freshwater lakes and streams that are being destroyed by acid rain. Instead of sulfur dioxide, oxides of nitrogen, a secondary component mainly produced by automobiles and electric utilities, is another culprit. Nitrogen oxides are converted into nitric acids and nitrates in the atmosphere. The damage to coastal aquatic life is the result of eutrophication, a greenhouselike effect in which excessive growth of algae, stimulated by nitrate salts and other nutrients, chokes off the oxygen supply and blocks the sunlight required by other plants and animals. According to the Environmental Defense Fund study, airborne nitrates issuing mostly from automobiles, power plants, and factories accounted for about one fourth of the nitrogen that is polluting the Atlantic Coast by fertilizing the excessive algae growth. Michael Oppenheimer, an atmospheric physicist and coauthor of the study, said that the study was significant because acid rain had previously been thought to affect only "a few hundred or a few thousand acidified lakes in remote areas." Now, he says, people will be really feeling its effects. "It is choking off the ocean's nurseries. It is shutting down the playground for millions of Americans. It is a regionwide problem, and it is only going to get worse because of the projected growth of nitrogen emissions."

More and more evidence is accumulating about the damage done by acid rain. A 1985 study prepared by the EPA, the Brookhaven National Laboratory, and the Army Corps of Engineers revealed that acid rain was responsible for an estimated $5 billion in damage to houses and other structures annually in an area comprising 17 states. The study surveyed 1,100 buildings in selected cities and also utilized previously existing data on pollution in a region that stretched south to Kentucky, west to Illinois, and northeast to Maine. The study's estimate of damage was based on the cost to repair or replace building materials damaged by acidic pollution. Signs of damage include breakdown of exterior paint (including chipping, blistering, and peeling), deterioration of roofing materials, and erosion of stonework on public monuments, statues, and decorative elements on buildings. What the report did not include in its damage estimate was at least as significant—automobile paint, roofing materials, the value of esthetic losses to cultural or historical structures, and, as Oppenheimer points out, it "did not include the costs of proved destruction of freshwater life and probably the much greater costs of damage to forests and public health."

The report also showed there was widespread damage not just in the states on the Eastern Seaboard but also in the Midwest. States like Ohio and Illinois, which are big emitters of the acidic pollutants, are also feeling the effects. Scientists now acknowledge that the problem affects all areas, not just a few high-altitude lakes in the Adirondacks.

Oppenheimer drew another important conclusion from the report: "Government estimates of the damage caused by acid rain to buildings and visibility were at least as high as the estimated costs of suggested acid-rain control programs." In fact, Oppenheimer estimated that "seven billion dollars was probably only a fraction of the costs to society caused by acid rain." He included an extra $2 billion in the cost figures to represent loss of visibility, which causes "loss of tourist income" and "increased delays and congestion at airports."

Another 1985 U.S. report also examined damage to forests; in this case, the Green Mountains of Vermont. According to scientists, in one area 50 percent of the red spruce trees at high elevations died in the last 20 years. One University of Vermont botanist who surveyed the region said, "It looked like a forest fire hit the area." The study's findings added to the mounting evidence indicating that pollutants can affect distant areas in the form of acid rain. The study's data showed the damage to be largely on western slopes that would be more vulnerable if, as some experts believe, the pollution is blown from as far as the Middle West.

In 1986, forestry experts began issuing reports that acid rain was threatening the sugar-maple trees in Vermont and Canada with extinction. One forestry researcher from McGill University says that industrial pollution is causing such a rapid decline in the number of trees, that "the delightful taste of maple syrup will remain but a memory." According to one maple-syrup producer, production has been dropping steadily—a 26 percent decrease from 1985 to 1986 in New York, a 38 percent drop in Vermont, and more than 50 percent in parts of Canada. Aerial surveys of the maple forests in Quebec began in 1982 and showed that about 32 percent of trees were seriously injured or dying. By the summer of 1986, the aerial studies showed that up to 82 percent of the trees were on the decline.

The decline of the sugar maple provides a clue as to how acid rain may be affecting all of our forests. Acid rain acts to weaken the trees, leaving them vulnerable to pests, diseases, and bad weather from which they are normally well protected. Parasitic insects called thrips spread rampantly through the maple groves during 1988, but there were signs that the trees were already soft before the thrips began to invade. "The big question," says sugar producer David Marvin, "is whether the pest problems are happening by themselves or whether opportunistic insects are attacking weakened trees."

A connection can be drawn between the infestation of thrips in the maple forests and other epidemic outbreaks that have recently occurred:

• Virus wiped out more than half of the harbor seals along North Sea coasts. Researchers believe it will continue to kill animals whose immune systems have been weakened by toxic chemicals.

• West German forest damage, probably from acid rain, rose to about 50 percent from 34 percent between 1983 and 1984, according to the Worldwatch Institute, a Washington, D.C., think tank.

• Researchers at North Carolina State University have found an epidemic aphid infestation of southeastern U.S. fir trees.

History and Politics: Lessons in Avoidance and Denial

Acid rain is part of a much larger problem, that of air pollution in general. Its solutions will only be found once the government begins to align its loyalties with the American public and the preservation of this wonderful country, instead of selling out to the powerful few who have all too long been calling the shots in Washington. The history of inaction on environmental issues by the Reagan administration was a tragedy for the nation and a disgrace to us internationally. We can only hope that the Bush administration will hold true to its stated commitment to the environment rather than auctioning it off to the highest bidder.

Why has there been so much foot-dragging on the part of politicians to clean up our environment, particularly when there is such a large body of evidence as to the damage being caused by pollution? Part of the answer to this question lies in what has been called the "Foul-Air Lobby," an army of lobbyists who for years have blocked legislation designed to control acid rain. Between 1980 and 1988, neither house of Congress voted on, much less approved, any pollution-control bills. One of the reasons for this conspicuous lack of action is the organized efforts of the industrial forces opposing clean-air legislation.

Among the leading lobbyists is Citizens for Sensible Control of Acid Rain, formed by a group of electric utilities and coal companies that has contributed $6.1 million to the effort.

The phenomenon known as *acid rain* has become one of the environment's worst blights. For almost 40 years scientists have issued reports on its devastating effects, and the public has watched as this by-product of the industrial age has destroyed forests and lakes throughout the world. Despite the controversy, very little has been done to implement solutions to the problem. In this regard, acid rain is no different than the multitude of other environmental problems we now face: The evidence is in, the public demands action, but the political factions continue to engage in specious debates designed to deny the all too obvious effects of the problem and to thwart initiatives that offer viable solutions.

What Is Acid Rain?

Acid rain results from certain kinds of air pollution that mix with precipitation, such as rain or fog, and then fall to Earth as an acidic solution. Its major components are oxides of sulfur and nitrogen that are primarily the by-products of coal-burning power plants, copper smelting, and factory and automobile emissions. These oxides are chemically altered in the atmosphere and return to the earth as rain, snow, fog, or dust. In the United States, the most popularly recognized form of acid rain results from sulfur-dioxide emissions, which are converted into sulfuric acid in the atmosphere. When this is mixed with precipitation and falls to Earth, the effect is precisely like pouring a diluted acid solution onto everything it touches. Although the effects of acid rain are most drastically felt and seen in the delicate ecostructures of our natural surroundings—the lakes, forests, farmlands, and oceans—its corrosive properties are now recognized to cause significant damage to even the man-made parts of our environment such as monuments and buildings.

In lakes, this acidification process can change ecological structures full of freshwater life into sterile, eutrophied dead zones. Particularly susceptible are those lakes with a low alkaline content or those that lack alkaline rocks and vegetation in their surroundings that can neutralize the acidic rainwater. Many of the nation's best trout-fishing spots have simply disappeared as a result of excess acidification. Once a lake has become too acid, bacteria and plankton die off, and little by little the entire food chain of the lake dwindles. Eventually frogs, fish, and other insects perish as well, leaving the lake itself barren and dead.

arena. The United States agreed with twenty-four other nations, including the Soviet Union, Canada, and most European countries, to freeze the rate of nitrogen-oxide emissions, starting in 1994, to levels no higher than those of 1987. The United States refused, however, to join 12 of the other signatories of the protocol in decreasing nitrogen oxide emissions by 30 percent over the next 10 years. The administration also refused to ratify an earlier protocol, endorsed by 16 of the participating nations in 1985, to decrease sulfur-dioxide emissions, claiming that the United States had already made significant inroads in cutting this type of pollution.

While the signing of the protocol is certainly a step in the right direction, one looming question still remains: Why is the United States in the rear rather than in the forefront of the effort to make this world a better place to live? The Canadian government, fed up with the foot-dragging of the EPA under the Reagan administration, followed the example of the northeastern states and took the EPA to court in April 1988. Before the U.S. Court of Appeals, the Canadian government petitioned for an order directing the EPA to enforce the Clean Air Act of 1970 and requiring American acid rain polluters to clean up after themselves.

Cause and effect are often difficult to determine with any certainty. One thing seems definite, though: The move to curb acid rain is both strengthening and making inroads. In June 1988, the wrangling and factionalism that have plagued the negotiations between the states for so long took a new and exciting turn: Governor Mario Cuomo of New York, a state considerably hard hit by acid rain, and Governor Richard Celeste of Ohio, one of the biggest sources of the pollution causing acid rain, signed a "peace plan" that proposed an annual line item to decrease toxic emissions and help industry pay for the cleanup. The plan was significant for several reasons. It showed that an accord is possible even between parties with very opposite interests. It came from the states, and not the federal government, and may be an indication not to count on the feds when it comes to environmental issues. Finally, it offered at least one positive and viable solution to the acid-rain problem. The proposal would cost about $1.8 billion annually, which would be spent to reduce pollutants and provide subsidies to industries for the cleanup. Most of the program would be funded by fees on polluters and on imported oil.

The New York Times heralded the plan as "an important break in the deadlock over Congressional legislation on the issue." But the *Times* admonished that "the deadlock over acid rain is only one element of a legislative stalemate that has persisted through the 1980s over proposals to amend and strengthen the Federal Clean Air Act."

Other encouraging news came in March 1989, when more than 100 members of the House of Representatives supported legislation that would roll back sulfur- and nitrogen-oxides emissions by 40 percent. Representative Gerry Sikorski, a Democrat of Minnesota and a supporter of the legislation, declared, "This is the year for clean air."

It is impossible to tell whether these moves toward problem solving and away from subterfuge and avoidance are due to the removal of the stumbling blocks set up by the Reagan administration during the past decade—Representative Sikorski cites the "new attitude" of the Bush administration as one of the factors facilitating the changes—or whether the outstanding efforts of Governors Cuomo and Celeste have finally given Congress the impetus it needed to proceed with much-overdue legislation, or whether the acid-rain problem has finally just reached that stage of critical mass in the conscience of America. Whatever the cause, for those of us who cherish and appreciate our environment, all we can say is that *it is about time!*

Suggestions for Individual Action

A key part of our personal health is the wellness of our environment. Unfortunately, for many years now Americans have been led to believe that there really isn't anything wrong with our environment. In fact, until very recently, the United States government's official policy on acid rain was: Why worry? What's the big deal?

We live in a very self-absorbed world today, a world in which people take care of themselves and don't bother about things that do not directly concern them. Someone living in the suburbs, for instance, can drive many miles to get to work in the morning, driving past hundreds of people's lives, and yet never feel a connection between anything he does and the effects it has on those people. We now know that virtually everything we do has some impact on our environment and on the people who surround us. We forget how intimately involved one part of our environment is with another, how upsetting one thing can in turn upset something else. We have assumed that we can just pollute and throw it away, and that by some magic it will all cleanse itself. Unfortunately, that simply is not the case. I was recently flying over parts of the United States, and I thought that there had been a forest fire that had burned out large sections of the underlying forests. Then I realized that that was not a forest fire, it was acid rain. And that sight is not uncommon.

There was a time when we were led to believe by government agencies, first, that acid rain did not exist, then, while it might exist, it is not a major problem; and finally, if it is a problem, we will have to study it for the next 20 years. Well, the time for studying is over, the time for action is now. Just what can we, as individual citizens, do to stop the ever-encroaching damage that is occurring daily as a result of acid rain?

Part of the problem with any environmental issue, and particularly with acid rain, is that the public is often bombarded with facts and statistics that are so overwhelming that it is unable to associate the problem with reality. What does it mean, for instance, when the OTA or *The New York Times* tells us that one quarter of the lakes in the northeastern United States are already damaged by acid rain? Fishing enthusiasts know what that means when it hits their favorite trout lake. Maple-sugar producers understand what acid rain means to maple forests, and hikers may know the damage being done to the nation's forests. But what does the typical American really know?

The first step in solving any problem is to become aware, *really* aware, of the problem. Acid rain is no exception. We need to experience what acid rain is doing to our environment by visiting those places where acid rain has hit hardest. Children can be made aware of the problem and encouraged to find solutions by taking field trips with their science classes to witness the devastating effects that acid rain is having on our environment. By educating our children on the risks associated with our excesses, the sacrifices they will have to make in order to set things right will not seem quite so onerous if they are associated with tangible benefits at an early age.

It is clear many of the solutions to the acid-rain problem are going to have to come from the grass roots and move upward. Although the Bush administration has promised to place top priority on environmental issues, the public needs to realize that the powers militating against any affirmative action on cleaning up our air and conserving our environment are politically very strong in Washington. For the moment, that is simply the way things are in our government. On the other hand, there are very real things that people can do, and one of them includes learning how to read a newspaper properly and intelligently, and how to listen to statements given on the issues.

Does former president Reagan's statement that we need no further controls because we have no proof of damage make sense to you? If so, then that is the end of it. After all, there are many faces to the truth, and *you* as an individual need to assess for yourself what is true for you. If the statement, or the probable consequences of the statement, do not make sense, then the next thing to do is to ask yourself what, if any, vested interest the person or agency may have in making the statement. It is up to each of you to determine whether you believe that this alignment of governmental statements and the vested interests of certain industries is merely a coincidence, or whether it indicates that further comments by the same governmental entities should be taken with a grain of salt. Remember, this is the age of information and communication. Whoever learns how to understand and assimilate information holds a great amount of power in this society today.

From a political point of view, one of the things that needs major modification in the acid-rain debate, as well as in all other environmental discussions, is what lawyers call "a shift in the burden of proof." As we have seen, both the administration and industry groups have vigorously adhered to the position that no further controls on toxic emissions are needed because there is no proof that they are causing any damage. Leaving aside for a moment the point that there is in fact ample proof of the damage caused by sulfur and nitrogen oxides, it should not be up to the public to have to prove, usually beyond a reasonable doubt, the causal link between the damage and the pollutant.

There is a need for a shift of consciousness among American legislators and policymakers. We know now that when a chemical or a drug or a pollutant is alleged to cause damage, more often than not, these allegations have some basis in fact. Most people, and scientists in particular, have better things to do with their time than to sit around dreaming up spurious claims of damage. It is only fair and reasonable to shift the burden of proof from its present position because of the vast amounts of resources companies, like the oil and coal industries, command in relation to most environmental groups, which are currently fighting our ecological battles on shoestring budgets and with volunteer help. Once there is a suspicion that a substance is harming our environment, then it is incumbent upon the producer of that substance to prove that the substance is not having the alleged effects. While industry should bear the financial burden, government should monitor commercial testing to assure validity of results.

For example, Exxon was applying for a permit to operate a natural-gas plant in Wyoming. One of Exxon's basic arguments in defending its permit application was that "there had not been enough research to indicate what level of acidity represented a threat to a lake." If this was in fact true, then instead of granting the permit, the Wyoming Environmental Quality Council should have required Exxon to prove that the sulfur dioxide it was adding to the environment was not going to result in damage to the surrounding lakes and forests. It makes absolutely no sense to allow a company to go ahead with plans to operate a plant, which will admittedly add to regional pollution, because the company alleges that there is insufficient research as to the level of pollution that causes environmental damage.

Each one of us can begin to think of taking responsibility in our personal and everyday life. There are a great many investment opportunities that are responding to the need for a cleaner environment. Companies engaged in activities like waste management or recycling are already starting to make large profits by cleaning up the environment rather than polluting it. We can encourage this shift in the economic power base by investing in these companies. This will make them stronger and more able to compete in the financial and political arenas with the powerful vested interests of the industries responsible for acid rain and air pollution. Capitalism can work as a system for all of us, not just a few megacorporations. If the bottom line is profit, then it is up to us to determine where the profits are going to be.

Investment is just one way to redirect the forces within our society. Changing our consumption patterns can be equally powerful. It is estimated that Americans are presently consuming 1 billion barrels of oil every 62 days. Every year, as a result of our combustion processes, we emit 26 million tons of sulphur dioxide into the atmosphere. We can begin to shift these consumption patterns and their toxic results by any number of consciously directed actions. Much can be done with respect to transportation: carpooling; pressuring local governments to provide more public transportation; using existing public transportation whenever possible; petitioning for auto-free zones and bicycle routes to encourage walking and cycling as alternatives to the automobile; buying more energy-efficient cars; and walking or cycling to work, which will not only cut down on pollution, but is also an easy way to exercise within the structure of your normal daily routine. Cutting down on the use of air conditioners and insisting that our workplaces do the same is of major importance in decreasing energy consumption. This will limit the emissions from power plants, and also will lower the amount of fluorocarbons released into the atmosphere.

Dr. Harriet Stubs, executive director of the Acid Rain Foundation offers some more suggestions for consumer action:

• Ask for paper bags in stores instead of plastic ones to take groceries home. Paper is recyclable, whereas most plastic is not.

• Do you buy juices in paper or plastic containers? Be aware of the packaging materials on the shelves and purchase the ones that can be recycled. Plastics are doubly harmful to the environment— fossil fuels are used to manufacture them, and they don't break down after disposal.

• Purchase drinks in returnable containers. The Wisconsin Department of Natural Resources has found that the cost to the consumer for returnable glass bottles is 21 cents, nonreturnable is 35 cents, for plastic bottles is 40 cents, and aluminum cans is 48 cents. Recycled glass bottles are cost effective.

• Conserve fossil fuels by lowering your thermostat to 65 degrees in the winter and increasing it to 72 degrees in the summer.

• Turn off your lights when you don't need them. These efforts will cut down on your utility bills too.

• Check to see if your car's catalytic converter is working properly, and also make sure the car is in tune. The better your car runs, the more efficiently it burns fuel.

These are just a few ideas to get you started on thinking about how you can make a difference in solving our environmental problems. Each person will have his own way of addressing the problem, all equally important to the final outcome. Additional information and educational materials for teachers and projects for students can be found at:

The Acid Rain Foundation
1410 Varsity Drive
Raleigh, NC 27606

Chapter 4
Water

Water, like air, is one of the basic elements of life. It makes up two thirds of our body mass, and while we can survive without food for two weeks or more, without water we would perish in a matter of days. But water is more to us than mere survival. It is our fun, our recreation and relaxation. Water draws us like a magnet whenever we want to unwind, get away and enjoy ourselves. Three quarters of our planet is covered with water. Just as we have assumed that our vast atmosphere could absorb, dilute, and detoxify whatever we dumped into the air, we have treated our seemingly boundless waters as if they had an unlimited capacity to wash away anything that we throw into them, be it tons of untreated sewage, solid garbage, or barrels full of toxic industrial wastes. For many years, this worked . . . or at least seemed to work. The waters are so vast, so deep, and flowed so quickly that they did in fact detoxify just about anything that we could throw into them. The "dilution solution to pollution" appeared to be a good one. So we dumped into the rivers, into the drains, and into the oceans, and because it all appeared to wash away, we thought that was the end of it.

The Long Arm of Pollution
The pollution of our waters is another example of society's short-sightedness and unwillingness to deal responsibly with its own waste. Although the problem is only recently receiving the front-page coverage it deserves, water pollution is almost as old as man himself. The sacred Ganges River is among the most revered of rivers and the most polluted. The Hindu religion views the Ganges as bringing spiritual cleanliness through daily immersion rituals. In the meantime, secular India dumps about 85 billion gallons of raw sewage into the sacred waters each year.

Although many Americans may not be aware of it, or prefer not to admit it, most of the cities in this country dump raw or inadequately treated sewage, solid garbage, and industrial and hospital wastes into our oceans and waterways in manners that are not all that different from those taking place on the banks of the Ganges. The only difference may be that we have long ago abandoned bathing and swimming in those waters because we realized the extent to which we had fouled them. Who in their right mind would consider swimming in the Hudson or East rivers that surround Manhattan, for instance? There once was a time when this would have been safe.

Since World War II, there has been a massive increase of chemical waste from the processing of steel, paper, textiles, food, electrical and transportation equipment, petroleum, and synthetics such as plastics, fibers, and detergents. During the war, the need for an abundant and cheap supply of goods led to the manufacture of synthetic alternatives, most of which were derived from petrochemicals, the by-products of oil refining. Since then, the petrochemical industry has expanded exponentially and is responsible for a major portion of today's chemical pollution. In 1940, for instance, some 25 million gallons of the solvent benzene were produced. Today, annual production has reached almost 2 billion gallons. Production of solvents as a whole increased 700 percent in the 15 years between 1968 and 1983, while plastics increased by 2,000 percent. Chemical production is estimated to contribute 60 percent of the country's hazardous waste.

According to EPA estimates, about 1,000 new chemicals are put on the market each year. Of the more than 50,000 chemicals now in existence, the agency deems about 35,000, or 70 percent, to present a danger to human health. We know dangerously little about the 12,000 potentially toxic chemicals now in wide use. The chemical industry itself provides no information. Hiding behind the veil of trade secrets, it tells authorities nothing about the chemical characteristics of the contaminants it contemptuously discharges into the nation's waters. We learn of the dangers only when they become manifest.

You can stand on the banks of the Potomac, in the center of our nation's capital, and watch the sewage ferment bubble up from the bottom of the river. The bed of the Potomac is composed of fourteen feet of odoriferous sludge. Its waters are green and brown: green from the rotting algae caused by massive eutrophication;

brown from the sludge, an amalgam of all varieties of waste and garbage dumped along the entire course of the river.

Following the course of the Potomac gives a true but sorry picture of the fate of today's waterways. At its very source, before the river even begins to flow, sulfuric acid seeps into the Potomac from working and abandoned coal mines. Below the coal-mining area, a pulp and paper mill pours its by-products into the river, mainly in the form of lignin, a fibrous substance that holds the trees together and is flushed out as a thick dark sludge in the manufacturing process. The mill is followed by the city of Cumberland, Maryland, which, after removing 30 percent of the solids from its sewage, releases the remainder directly into the Potomac. After Cumberland comes a series of textile factories that add vatloads of poisonous dyes. A tire factory and a plate-glass-window factory then add their effluents to the mixture. At Berkley Springs, a sand mine contributes a choking mass of sediment—one of the worst types of pollution. By the time the city of Washington gets to unload its share of sewage, garbage, and chemicals into the river, it's kicking a dead horse.

Many Americans have become so nonplussed about the state of our waters that often very serious situations are ignored. In late 1983, a cancer epidemic affecting 100 percent of the inhabitants of a site in Michigan slipped in and out of the news without arousing much notice at all. Granted, the site was the little-known Torch Lake and the inhabitants were the goggled-eyed sauger pike, but any such epidemic, even among fish, is an indication that something is seriously wrong in the environment. In five widely distributed locations in the United States, an epidemic of skin and liver cancer has hit freshwater fish living in proximity to polluted industrial areas. "In each of these scenarios," reported Congressman John Breaux, "a known, man-generated carcinogen has been identified as the suspected cause of these massive outbreaks of cancer." Scientists testifying at the congressional hearings maintained that the examples were part of a larger, more dangerous problem.

While the fate of a fish population in a remote lake is not apt to evoke the same sympathy as a human epidemic like AIDS, any unusual outbreak of cancer cannot be lightly dismissed. Scientists are becoming increasingly aware that animals and fish are often among the first indicators of imbalances in our ecosystem—imbalances that will eventually take their toll on human life as well. Mercury poisoning in Japan was signaled by sick cats, the fatal deficiency of oxygen in Lake Erie was first noticed by the disappearance of mayflies, and the discovery of dioxin in the Great Lakes followed the attention given to eggshells that were so soft they could be dented by the gentlest handling. Fish, with their sensitive physiologies, are considered ideal sentinels of cancer-causing conditions. As suspected, when scientists investigated the areas surrounding the five sites in which the fish were dying of cancer, they also found an unusually high incidence of human cancer.

For many generations, the Mohawk Indians lived in harmony with their environment on the shores of the St. Lawrence River. With the completion of the St. Lawrence Seaway in 1954, industries began to migrate to shores of the great river. Shortly thereafter, Native Americans on the St. Regis Mohawk Reservation began to experience health problems they had never heard of before. When biologist Ward Stone of New York State's Department of Environmental Conservation began working with the Mohawks to investigate their suspicions that pollution was infiltrating their waters and causing their illnesses, his discoveries came not from the examination of the Indians themselves, but from a fourteen-year-old snapping turtle—laden with toxic chemicals. Stone had captured the turtle within 300 feet of a corporate landfill, a site the Mohawks had long suspected as being a source of pollution. An analysis of the turtle's fat tissue revealed polychlorinated biphenals (PCBs) at an alarming rate of 835 parts per million (ppm) as well as extremely high levels of dioxin. According to Stone, one ppm of PCBs will cause reproductive failure and death in minks, while two pounds of PCBs distributed through a million pounds of fish will make the fish unsafe for human consumption. While companies on both the U.S. and Canadian sides of the border denied that they were dumping any pollutants into the St. Lawrence, further study of the river's fish and frogs eventually led to tracking many of the sources of pollution. Stone found frogs that did not behave normally—they ended up having as much as 2,000 ppm of PCBs in lipid areas of their bodies. Local plants , which initially denied any emissions of PCBs, were found not only to be leaking large quantities of the substance into the surrounding water, but were also found to be a source of fluoride air emissions.

James Ransom, director of environmental programs for the St. Regis Reservation, had first observed the fluoride contamination when cattle began to die from clinical fluorosis. "They eat the grass, and their teeth become brittle and fall out, " he said. He also noted the existence of birth defects. Local fishermen related stories, and

Stone has a collection to confirm them, of fish with no skin, or deformed fish with tumors or no backbone. This led researchers to investigate industries along the river and to find out that large international companies were discharging large quantities of mercury into the water. Today, the landfill has been declared an EPA Superfund site. Says Ransom, "They've identified about 350,000 cubic yards of PCB-contaminated waste materials and soils on their site. We think that it is probably the biggest PCB-disposal site in the United States."

It has been many years now since we have seen that dilution will not solve pollution in our lakes and streams. While it still may be possible to find patches of clean air, there are very few large bodies of clean water left in the United States. As early as 1953, Lake Erie was so contaminated that even mayflies could not hatch on its surface. The lake is so polluted that its beaches have long been declared unsafe. Although many citizen and environmental groups have made significant inroads in cleaning up the Great Lakes, their efforts are a constant uphill battle. The EPA only sets standards for a very few chemicals, and the chemical industry is very creative in its ability to come up with new substances not covered by federal regulations. Furthermore, we are only just beginning to get a handhold on the effects these chemicals have not simply on human health, but also on the wildlife inhabiting the bodies of water where dumping has taken place. A 1987 study commissioned by the Michigan Department of Natural Resources revealed birth defects and reproductive failures in a number of species of birds feeding on fish in the Great Lakes. According to James Ludwig, a wildlife specialist who conducted the study, these findings suggest that toxic threats to humans in the area may have been overlooked. Ludwig believes the state puts too much emphasis on levels of toxins, and not enough on their effects.

In June 1968, the Cuyahoga River, which runs south from Lake Erie at Cleveland, burst into flames. Thousands of gallons of oil—the source of which has never been determined—suddenly caught fire, causing extensive environmental damage. For years preceding the fire, the ugly and polluted Cuyahoga had been a joke in Cleveland. Local radio disc jockeys did regular comedy routines about the river. When it caught fire, the ultimate punch line was delivered.

It is no joke that other rivers, such as the Buffalo, the Grand Calumet, and the Mahoning, are so slick with oil and chemicals that they are considered serious fire hazards. Rivers that may not be fire hazards are so contaminated by chemicals, pesticides, domestic garbage, and industrial effluents that they are dead or dying and dangerous to any forms of life using their waters. Two studies by the EPA, the results of which were made public in March 1989, confirmed many people's fears about both the amounts and kinds of chemicals being poured into our waterways by the country's industries. The studies focused on the dioxin emissions of paper mills. They found fish downstream from 21 of the 81 mills examined contained levels of the chemical far surpassing those set by federal authorities as hazardous. A second study revealed that although levels of dioxin in 59 of 74 mills visited were low, they were still far above levels set by federal standards under the Clean Water Act of 1972.

Dioxin is a by-product of the chlorine used to bleach paper white. It is also a component of Agent Orange, and a recognized carcinogen that has also been linked with immune-system damage and a skin condition called chloracne. While scientists still dispute the amounts that will cause cancer and disease in humans, dioxin is known to be extremely toxic. Fish downstream from the paper mills were found to contain up to 7 times the 25 parts per trillion level set by the FDA as being dangerous for human consumption.

While paper mills are not the only riverside polluters, they are among the largest and also provide a good example of the type of politics large companies engage in to thwart pollution controls. One giant paper mill located in North Carolina has to contaminate 45 million gallons of river water each day in order to maintain its normal production levels. During the summer, the flow of the river, not voluminous to begin with, trickles down to 50 gallons a day. The paper company solves this problem by simply channeling the entire river through its mill. The entire town has been built around the mammoth factory. Residents working at the factory earn an average of $14 an hour and often make in excess of $30,000 per year, compared with a statewide annual wage of $13,000. The town smells of the paper mill—a strong rotten-egg odor that permeates indoor and outdoor areas alike—and the river runs from yellowish-brown to a thick dark coffee color from the sludge discharged at the mills. The townspeople do not seem to mind either. Unfortunately, the odor translates into income; one union leader says, "It smells like money to us."

Polluting industries often use the argument that cleaning up after themselves will cost the nation jobs. Management convinced the paperworkers of exactly this—that without pollution there would be no jobs. The workers fought by the company's side to help it avoid having to install costly new pollution controls, but ironically the workers lost their jobs anyway. Management did not intend to make idle threats. Consequently, notwithstanding workers standing by the company, when Tennessee citizens living some 40 miles downriver from the paper mill put pressure on their state government to refuse to grant a variance to the company that would have allowed it to continue dumping its effluent into the nearby river in quantities exceeding Tennessee's clean-water standards, the corporation used this as a pretext for firing 1,000 workers and said that in order to meet the Tennessee standards, it would have to shut down four of the mill's six papermaking machines. Industries surrounding the Great Lakes, on the St. Lawrence, the Mississippi, and around the world are waging the same battle. They are trying to convince resident workers and politicians that without business as usual, there will be no jobs and no industry. Understandably, this is very scary for workers who depend upon the industry for their livelihood, but it is also utter nonsense. Companies have always passed on costs to consumers in order to maintain their target profit levels. If an industry is taxed or has to pay for its pollution, it will simply add the costs on to the final product. Moreover, even in the case of this North Carolina paper mill, where the company faces a deadline for pollution cutbacks, if the company is making good-faith efforts to comply, it is hard to believe that an extension of the deadline could not be negotiated, particularly if a large number of jobs are at stake.

One of the most serious and tragic aspects of water pollution is a biological process called eutrophication. This occurs when a body of water becomes abnormally rich in certain nutrients that feed excessive algae growth on the water's surface. This draws oxygen from the water and causes aquatic life to die. The primary causes of pollution-related eutrophication include phosphates and nitrogen compounds entering the water from agricultural runoff, acid rain, and sewage systems. Unless quick and effective measures are taken to address the eutrophication, once it sets in, it usually sounds the death knell for the affected lake or stream. In mid-summer of 1986, a massive bloom of blue-green algae erupted on 100 square miles of Lake Okeechobee fueled already existing alarms about the imminent death of the lake. Often called the "liquid heart" of South Florida, the 720-square-mile lake is the core of the Everglades and provides a reservoir for the coastal cities of Miami and Fort Lauderdale.

Water-district officials believe that the loss of Lake Okeechobee would be an environmental disaster. Unless Florida adopts a more responsible approach to water pollution than it has in the past, another of the world's invaluable water resources will simply cease to exist. This is not the first freshwater lake in Florida to die of asphyxiation due to excess nutrients. In the 1960s state officials just sat back and watched while Lake Apopka in Central Florida died. Studies on how to stop the deterioration of Lake Okeechobee had been prepared by environmental groups as early as 1976, but were ignored for the most part. For decades the South Florida Water Management District's board of governors was principally concerned with flood control and was perceived to be under the influence of the agricultural interests it served, especially the sugarcane growers south and east of the lake. A procedure in which water is pumped from the waterways crisscrossing the sprawling cane fields back into the lake for storage continues despite evidence suggesting that it gorges the lake with nutrients.

In March 1989, more trouble hit Florida's waters. This time it took the form of mysteriously high levels of mercury appearing in game fish from the Everglades, causing state officials to warn against eating largemouth bass and warmouth caught in the region. According to Frank D'Itri, a mercury expert at Michigan State University, the fish contained one of the highest concentrations ever recorded in the country. A number of theories were advanced about the source of the mercury. They were all equally dire in portending an even further decline in the fragile ecosystem of the Everglades. Could it have been midnight dumpers who smuggled the load into the Everglades from an electronics plant, for instance? Or had the mercury been in sludge of the lake for years, perhaps from the days when mercury was still allowed for use in pesticides? Was it now coming to the surface, perhaps some 20 years later, because drought and water shortages (probably fueled by other environmental problems) caused the water authorities to draw more heavily upon the water resources of the Everglades? While no one knows for sure yet, the deterioration of the Everglades and the Florida water system is certain to continue unabated unless something is done very soon.

The Death of Our Oceans, Beaches, and Marine Life
The oceans and beaches of America are not faring any better than its freshwater streams. The summer of 1988 brought all-time rec-

ord waves of garbage, hospital waste, and raw sewage onto our nation's beaches, causing many of them to close. A 1987 federal study
revealed that harbors around the country are filled with pesticides
from agricultural runoff, industrial chemicals, and sewage that
concentrate and collect to pollute the sediment. Among the toxic
accumulation are pesticides such as the banned DDT and known
carcinogens like PCBs. The study also revealed significant levels of
these toxic chemicals turning up in the groundwater supplies of
harbor cities. Levels of pollution were found to be the worst in the
harbors of Salem and Boston, Massachusetts, and Raritan Bay between New York and New Jersey, but the rest of the country's harbors still have unacceptably high levels of contamination. The
highest levels of DDT in the country were found in California's San
Pedro Canyon, the result of years of dumping industrial waste into
the Los Angeles harbor by a DDT manufacturer.

Similar toxins are polluting seas and causing beaches to close
around the world. Sylt, an island in the North Sea long cherished
by European vacationers for its pristine beaches and beautiful
shores, was struck by an unexpected wave of pollution. Fish on the
eastern shore of the island, just south of the Danish border, were
found with tumors and lesions. Scientists believe the epidemic to
be caused by industrial contamination dumped into the Rhine,
Elbe, and other rivers connected to the North Sea. The once-im-
maculate island was also the first place where seals died during an
epidemic in which 7,000, roughly one half, of the North Sea's population of the mammals were killed. Again scientists suspect pollution; this time it acted on the seals' immune systems and caused
them to succumb to secondary infections, much as AIDS works on
humans.

As with our bodies of fresh water, ocean-dwelling life is also
showing increasing signs of ill effects from pollution. Man's disregard for his oceans has made victims out of the marine animals
whose home it is. It is one thing if we swim in our own waste—after
all, we created it. We have a choice whether or not to bathe in particularly polluted places, but for the animals, this is their habitat,
they have no alternative. Particularly hard hit have been the
beaches on the East Coast of the United States. During the summer
of 1987, low-oxygen conditions caused huge numbers of lobsters to
die all along Long Island Sound, while dolphins washed ashore by
the hundreds from New Jersey to Florida. Widespread algae
blooms, often called "brown tide," making waters murky and
choking off marine life, plagued the East Coast in both 1987 and

1988. In July 1988, New Jersey's beaches were awash with tens of thousands of dead fish—eels, flounder, bass, shrimp, and crab— sparking even greater fears among scientists about the decline in the region's waters.

As we continue to mistake the vastness of our seas for a limitless dump for human waste of every imaginable description, we not only recklessly foul one of our most valuable resources, but are also directly responsible for the death of thousands of marine animals. At one time this sort of mass dumping did not create such a problem. For the most part, the waste, such as paper, food, sewage, or cloth, was biodegradable. Metals and glass would sink to the bottom of the ocean and disappear. The emergence and proliferation of plastics have created a whole new generation of environmental problems. Plastics are light, so they float when they are dumped into the ocean; they do not biodegrade, so for years they float on the water's surface, causing unsightly debris entanglements. Additionally, as columnist Robert Walters points out, these materials are also responsible for great harm to our sea life:

•Sea turtles regularly mistake plastic bags for the jellyfish that are their favorite food. They die because they can neither ingest anything else nor expel the poisonous material that clogs their digestive systems.

•Entanglement in six-pack holders and other plastics has contributed to an annual decline of 4 to 8 percent in the fur-seal population of Alaska's Pribilof Islands.

•In the digestive tract of a single sea turtle found dead on a beach in Hawaii, medical researchers found a golf tee, a pocket comb, a plastic flower, pieces of monofilament fishing line, part of a bottle cap, an eight-inch square plastic bag, and dozens of other pieces of plastic.

Where does all this plastic come from, and is it really such a problem? The world's fleet of more than 70,000 merchant vessels dumps at least 450,000 plastic, 300,000 glass, and 4.8 million metal containers into the sea every day. This country's commercial fishing fleet of 130,000 ships yearly discards 245 tons of nets, ropes, traps, buoys, packaging material, and other plastic items into the ocean. More plastic garbage is tossed overboard from the navy's 600 vessels and from cruise ships, recreational boats, and offshore oil rigs.

Throughout the summer of 1988, New York and New Jersey newspapers were running daily stories about the ever more visible results of pollution. The severity of the problems off the Jersey shore invited even tabloid coverage, and for a change, the blunt sensationalism did not seem out of place. At the height of the summer, *The New York Post*, one of the most popular tabloids, did a story on a crusty sea captain and his observations of the changes that had taken place in areas around the sewage discharge pipes off the coast of New Jersey.

"Twenty years ago," says Charlie Stratton, a veteran seaman interviewed by the *Post*, "dolphins were jumping. We had sea turtles, rainfish, mackerel, all sorts of tuna. Now there's not even jellyfish."

Today, this is a dead sea.

New Jersey has passed legislation ending the dumping of sewage sludge off its coast by 1991, and also calling for increased surveillance of illegal ocean dumping and five-year prison sentences for violators. The long-overdue law is welcome, but many fear that without a more comprehensive approach to ocean pollution, the law stands little chance of successfully dealing with the problem.

The Sewage Dilemma

The dumping of raw sewage into the world's rivers and oceans occurs almost everywhere that you find civilization. Around the Mediterranean, 70 percent of the coastal cities dump untreated sewage directly into the sea. As recently as 15 years ago, New York City's treatment plants discharged 450 million gallons of raw sewage a day into New York Harbor, the Hudson River, and other local waters. Since that time the city has invested $2.5 billion in treatment plants that city officials claim have limited the raw sewage to 1 million gallons a day, but this only begins to dent the problems of sewage pollution in the area.

There is also the problem of sewer-system design and the questionable efficiency of current treatment systems to remove polluting materials from the waste adequately. In New York, New Jersey, and Connecticut, for instance, many sewer systems are combined with storm-drainage networks. According to R. Lawrence Swanson, director of the Waste Management Institute at the State University of New York at Stony Brook, with this system, "It takes a rainfall of only a few hundredths of an inch per hour for an hour or more in the metropolitan area for sewage treatment plants to be bypassed." What this means is that with very little rain, sewer networks become backed up, and rain-diluted sewage, instead of ending up at the treatment facility, is dumped in an untreated state directly into the area's waterways. Municipal estimates for New York City put the level of sewage overflow at about 65 billion gallons per year.

Part of New York City's problem stems from a sewer system dating from the early 1880s that was never designed to deal with the number of people now living in the city. While New York's problem may be more serious than that of other regions, rapid development and poor city planning, particularly in coastal regions, make this a nationwide problem. People do not like to think about waste and sewage. Developers do not want to pay for the new systems that are realistically required with the construction of new communities. Municipal-planning officials often give way to developers' demands, not wanting to squabble about waste. When problems become apparent, the price tag of remedying the situation is so exorbitant that taxpayers either cannot or do not want to spend the money to build a new infrastructure.

New York City is just one case in point. For quite a few years now developers have been allowed almost carte blanche in the construction of high-density residential and office buildings. If they have paid at all for rebuilding the city's aging and overtaxed infrastructure, it has been minimally. Environmentalists believe that to address the problem responsibly, a new system must be built that separates sewage from drainage water. City officials dismiss this out-of-hand as too complicated and costly. Instead, high-priced stopgap plans that are almost absurd because of their shortsightedness are proposed and adop-ted. The city currently has a 10 - year, $1.2 billion sewer-rehabilitation program on the table. Under consideration is a series of underground tanks to hold sewage and drain water following storms. Screens would be placed over sewage -discharge points to prevent solid matter from entering the waterways.

The simplemindedness of this approach is understandable given the manner in which sewage issues have been handled up until now. The $2.5 billion spent by New York City during the past 15 years for treatment plants address only a fraction of the problems caused by sewage. Not only are sewer networks insufficient to handle the city's volume of waste, the sewage reaching the treatment facilities hardly leaves those facilities in a clean state. Most sewage-treatment plants, in this country and around the world, only remove and treat the solid portions of the waste. Once treated, it is released into the waterways and oceans as a thick dark sludge. Liquid waste is allowed to pass through these plants without any alteration. These liquids are filled with all types of pollutants, from bacteria and industrial waste to agricultural runoff filled with eutrophying nitrates and phosphates. It is this portion of our sewage waste that is believed to be the primary culprit in the widespread algae blooms that are choking off so many of the world's coastal areas and causing such widespread devastation to marine life.

Oil Spills

Just six months after the end of the disastrous summer of 1988, America watched in horror as the *Exxon Valdez* emptied 10 million gallons of crude oil into the waters around Prince William Sound in Alaska—fouling, perhaps irreparably, 800 miles of what had been one of the world's most unspoiled and picturesque coastlines. The accident, the largest oil spill in U.S. history, followed closely on the heels of two other major oil disasters. In December 1988, Grays Harbor in Washington State was the site of a 300,000 gallon accidental spill. On January 28, 1989, a calamity began on the opposite pole when the Argentine ship *Bahia Paraiso* struck a reef and began leaking 3,000 gallons of fuel and crude oil a day into the previously pristine waters surrounding the Antarctic Peninsula.

The magnitude of these disasters and their rapid succession sent shock waves through the American public, which for years had placed blind trust in oil-company assurances that such things could never happen. But they did, and particularly in the wake of the *Exxon Valdez* spill in Alaska, investigation into oil-company precautions revealed a cavalier approach to safety.

As far back as the 1970s , Exxon and the other companies that own the Alaska pipeline assured environmentalists that they had a cleanup plan that could contain a major spill with five hours of a rupture. But in 1981 the industry disbanded a 20-member emergency team prepared for round-the-clock responses to oil spills.

In the case of the *Valdez,* Exxon's cleanup equipment was totally inadequate, giving the leaking oil two full days to spread before anything significant was done. This lack of preparation and foresight was confirmed on television and press interviews with Exxon Chairman L. G. Rawl, who made the shocking admission that the Alyeska Pipeline Service Company, a consortium of oil companies, including Exxon, responsible for spill containment, had never been prepared to handle a spill of this magnitude.

Alaska state officials, environmentalists, and fishermen have believed since the early seventies that a tanker accident in the sound was inevitable. To combat these concerns, Alyeska and federal officials promised in 1972 that the tanker fleet operating out of Valdez would be specially designed to minimize spills by incorporating such safety features as double bottoms and protective ballast tanks.

By 1977, however, Alyeska had persuaded the coast guard that the safety features were not necessary, so most ships, including the *Exxon Valdez*, were not outfitted. At the time of the accident, Alyeska's containment team consisted of a single barge loaded with 7,000 feet of barrier booms. With the *Valdez* measuring almost 1,000 feet itself, the 7,000 feet of booms readily deployable could barely encircle the giant ship, much less a sizable slick.

Evidence such as this left many infuriating questions unanswered, the most obvious being that if oil companies are not prepared to handle a spill of this magnitude, or anything close to it, why did they not limit their shipments to cargos that could be handled if spilled?

Perhaps we all would have been more sympathetic to Exxon had the spill been the result of a *real* accident—what lawyers call *force majeure* or "acts of God," like a totally unforeseeable storm or even an uncontrollable equipment failure. But the Exxon accident brought to our attention a picture of corporate mismanagement and recklessness. What else could explain a drunken captain, with a history of alcohol abuse, driving a 978-foot, cargo-laden oil tanker onto a shallow reef?

The Exxon oil disaster could hardly have occurred at a worse place. Throughout the media coverage, the pristine beauty of Prince William Sound was stressed. The postcard view of evergreens and mountains against a dazzling blue sky featured prominently in all the reports. Crude oil, carried in by Arctic tides, left a 12-foot-wide brown carpet along the desolate beaches.

The toll on animal life in the area was enormous. At least 82 sea otters were brought to a makeshift field hospital in Valdez. They were nearly frozen because a coat of oil had destroyed the insulating ability of their fur. Animals dead on arrival steadily filled up a white refrigerated truck trailer parked nearby. A preliminary beach survey indicated an average of 80 oil-coated ducks and other kinds of birds every 100 meters. Bald eagles have scavenged the contaminated birds, and they are now at risk.

Crude oil contains substances that are either poisonous or carcinogenic. The danger from contaminated fish prompted state officials to announce that this year's herring season, expected to bring fishermen $12 million in revenues, would be canceled. Salmon fisheries are also in danger.

Studies by the National Academy of Sciences and other groups have found that the largest single source of oil pollution in ocean regions is crude oil tankers discharging tank flushings and ballast. In the North Sea alone, it is estimated that 400,000 tons of oil are released into the water each year from offshore oil rigs and ships emptying their tanks at sea. An investigation of petroleum pollution in the Gulf of Mexico has revealed the Gulf to be among the most polluted major bodies of water in the world. A Florida survey team said that oil from blowouts of offshore wells such as the Ixtoc 1, which spilled 134 million gallons of oil into the Gulf in 1979, account for only a small percentage of oil identified in the region.

Tank flushings and ballast-water discharges by tankers taking on crude oil in Alaska are adding 13 million gallons of contaminated water to the sea each day. A water-treatment plant designed to remove oil and toxic pollutants from the tankers' discharges turned out to be a fraud. The company maintained no pollution-control monitors, and its managers gave false testimony about potential hazards and tampered with environmental-impact statements to make the company appear to be doing a better job than it was.

Though our energy-hungry world needs petroleum, and America needs its domestic sources, we need to go after our fuel in environmentally responsible ways, and move toward alternative forms of energy. It is only through decreasing our fuel consumption that we can slow down oil drilling and preserve our oceans and beaches. We must also push our legislators to take a long-term approach to all issues relating to oil, be they conservation, drilling, or dumping.

Drinking Water
Up until recently, an abundant supply of safe drinking water has been taken for granted by most people. Today, hazardous-waste sites are leaking and contaminating the underground aquifers that supply more than 53 percent of the nation's drinking water, including 97 percent in rural areas.

The following are a few other alarming facts about the safety of our drinking water:
• A 1982 EPA survey of large public water systems supplied by underground aquifers found 45 percent of them to be contaminated with organic chemicals;
• In New Jersey, every major aquifer is contaminated by chemical pollutants;

• In California, pesticides contaminate the drinking water of more than 1 million people;

• On Long Island, more than 4,000 wells are contaminated with Temik, a potent pesticide used in the region for potato farming;

• Underground aquifers surrounding almost all of the nation's nuclear-weapon facilities have been contaminated with radioactive waste and other toxic chemicals. Some of the most severely affected areas include those surrounding Fernald, Ohio; the Mound Facility near Miamiburg, Ohio; Livermore, California; and Hanford Reservation near Richland, Washington. In Texas, the Pantex Facility, has not established sufficient safeguards to protect the Ogallala Aquifer, the main source of drinking water for Amarillo, from its waste. The Savannah River Plant near Aitken, South Carolina, has contaminated that area's most important aquifer.

Of the more than 48,000 chemicals listed by the EPA, little or nothing is know about 35,000 of them. Less than 1,000 of them have been tested at all for their health effects on humans, while only about half that number have been analyzed for their ability to produce cancer. Despite the real or suspected danger of many of these chemicals, the EPA regulates levels for only 30 of them in your drinking water; and often that regulation stems from years of citizen action rather than EPA diligence to protect public health. The agency has also proposed to stop setting zero-contamination goals for chemicals suspected of causing cancer. Standards for compounds in which a link to cancer is not "clearly established" would become even less stringent. Congress had directed the EPA to increase the number of regulated chemicals to 83 by June 1989, to publish a list of suspected contaminants, to adopt regulations concerning 25 of those substances by 1991, and to regularly adopt standards for 3 more chemicals each year thereafter. As usual, however, the EPA was slow to act. In the meantime, according to a report released by the National Wildlife Federation in October 1988, "millions of Americans are drinking unsafe water." The federation blamed the federal government for failing to enforce the Safe Drinking Water Act. Upon review of federal documents the wildlife federation, the nation's largest environmental group, found that in 1987, public water systems serving 37 million people violated the law 101,588 times, but that states took action on these violations only 2,544 times and the EPA acted on only 50 cases.

The federation's report echoed an earlier study conducted by a group affiliated with consumer advocate Ralph Nader that condemned EPA inaction and criticized the agency for being derelict in its duty. The group, Center for Responsive Law, found that nearly one out of five public water systems across the country were contaminated by chemicals.

The following are some of the contaminants, how they commonly find their way into our water supplies, and how can they affect our health:

Benzene: A clear, colorless liquid derived from petroleum. It is used in the manufacture of pesticides, detergents, pharmaceuticals, paints, plastics, and motor fuels, and is widely used by industries as a solvent. Benzene is a potent carcinogen. It enters our water supply primarily as industrial waste, and agricultural run-off, and through leaky fuel tanks.

Some 40 percent of the nation's groundwater may be contaminated by gasoline the complex chemical structure of which contains not only benzene, but also many other dangerous substances like ethylene dibromide (EDB). One gallon of gasoline can contaminate 75,000 gallons of water with these toxic chemicals. The EPA estimates that nearly a quarter of the 2.5 million gasoline storage tanks across the nation may be leaking. Most of these tanks have already outlived their normal life expectancy, but the oil companies have been slow to replace them. In some contamination situations, the companies have compensated people whose water supply was contaminated by the leaks, but they are settling these matters out of court and hence have not made any admission of guilt. Around Denver, for instance, Chevron bought up 41 homes and relocated residents at a cost of $10 million when groundwater was found to be contaminated by gasoline leaks. Mobil and Exxon spent $1 million to build a new water system in Canobe Park, Rhode Island, when existing water supplies became tainted by gasoline.

Trichloroethylene: Also known as TCE, this chemical has accumulated in the environment from the disposal of dry-cleaning materials, the manufacture of pesticides, paints, waxes, varnishes, paint stripper, and metal degreasers. Minute amounts of this chemical are suspected of causing cancer. For residents of Des Moines, Iowa, levels of this deadly chemical in their drinking water climbed to more than 18 times the federal standard of 50 parts per billion in 1983. The cause: A local company had used the chemical in the sixties as a solvent to remove grease from brakes and wheels. When it came to disposing of the stuff, the company sprayed it on its parking lot to control dust and dumped the rest down the drain.

In Fort Edwards, New York, it was only after TCE reached levels of 2,000 times the federal standard that officials from the New York

State Board of Health told residents that they should discontinue using their water. By this time, levels were so high that health officials advised residents to simply shut off their taps. Showers were dangerous because TCE can vaporize and be inhaled. If residents flushed the toilet, they were supposed to shut the lid and get out of the bathroom. A suspected source of Fort Edward's problems was a General Electric factory that leaked and soaked the ground with TCE. Rains drove the chemical further into the earth, where it contaminated the city's groundwater.

Inorganic Chemicals:

Arsenic is known to cause skin and nervous-system disorders. It enters our water supply through pesticide residues, industrial waste, and smelting operations. The Industriplex-128 industrial park in Woburn, Massachusetts, has been found to contain huge arsenic pits dating back from 1863, when the Merrimac Chemical Company moved to the site and began manufacturing arsenical insecticides. The complex is now designated as one of the 40 Superfund cleanup sites across the nation.

Barium can affect the circulatory system. It enters our water as a result of pesticide residues, industrial waste, and smelting operations. A 1983 EPA report found that 1,500 to 3,000 municipal water systems across the country rely on groundwater that contains excess levels of arsenic, barium, and lead.

Cadmium and *Chromium* mainly come to us from geological mining and smelting. Both affect the kidneys, and chromium is toxic to the liver. The Industriplex has also been found to contain huge chromium lagoons developed by glue manufacturers who used chromium-treated tannery scraps. The chromium, as well as the arsenic and other chemical deposits, are now leaching and contaminating the area's groundwater supplies.

Lead is one of the most prevalent and most toxic forms of water pollution. It is extremely toxic to children and pregnant women, and can result in delayed physical and mental development in babies and mental impairment in children. Excessive lead can also lead to nervous-system damage, hearing loss, anemia, and kidney damage. Even in small amounts the metal can inhibit red blood-cell formation and cause low-weight births. Lead usually enters our water by leaching out of lead pipes and lead solder pipe joints.

The Centers for Disease Control (CDC) estimates that 10.4 million children are exposed to excessive amounts of lead in their drinking water. According to EPA senior scientist Joel Schwartz, "The more we learn about lead, the more we find adverse effects at lower and lower levels. Drinking water is now a major source of lead for a sizable portion of the population."

To express its concern about high lead concentrations in America's drinking water, the EPA has issued proposals to decrease levels of the metal, but most environmentalists view the proposals as weak and ineffectual. First, they address the lead problem by recommending that water be chemically treated to decrease its ability to leach the metal from pipe. With the current level of chemicals in our water supply, this chemical solution is a poor response to the contamination of our water. Moreover, the EPA proposals do not set numerical limits for the amount of lead that can be found in water coming from public systems, but instead only tell water authorities to use their "best efforts" to decrease quantities to the lowest possible level as determined by state and federal authorities. In the event that the water does not meet these standards, the water suppliers do not have to take any further action except to conduct public education programs to show people how to decrease their exposure.

Public education programs are important, particularly in poorer areas where people may have old and corroded pipes contaminating their water, yet they cannot afford bottled water. But they are certainly not a substitute for decisive regulation and control. Nevertheless, until the EPA and other governmental agencies decide to act more responsibly, you need to be informed about the measures you can take to reduce exposure to lead. The ideal solution is to replace all pipes made of lead or containing lead solderings. For most people this is not practicable because replacement is too expensive. One thing that everyone can do, however, is to let water run for a few minutes before using it for cooking or drinking. This allows for the clearing of water that has been sitting in pipes and accumulating lead contamination.

Particularly alarming because of lead's toxicity to children was a finding from a congressional-investigation panel that many, if not most, of the electric water fountains used in schools and buildings were releasing unhealthy levels of lead. Paul Mushak, a professor of environmental pathology and coauthor of a study on lead poisoning in children, testified before a congressional subcommittee that the lead was entering the drinking-fountain water through lead-lined storage tanks and lead soldering used in their construction. He warned that the fountains, often found in schools and day-care centers, posed very serious risks to children because of their high susceptibility to lead poisoning.

Fluoride: For nearly 40 years, respected organizations like the American Dental Association and the U.S. Public Health Service have promoted the claim that fluoridation prevents dental cavities. Responding to these claims, cities around the country have added this chemical to their drinking-water supplies, usually at levels of about one part per million.

There are no published studies showing that fluoride reduces tooth decay. What we do know is that excessive fluoride causes skeletal damage, and a disorder called fluorosis in which teeth develop white spots and become brittle. There is also mounting evidence that fluoride causes cancer. According to Dean Burk, a senior researcher at the National Cancer Institute, in the United States alone more people have died in the last 30 years from cancer connected with fluoridation than all the military deaths in the entire history of the United States.

While fluoride does appear naturally in water, sometimes even in quite elevated levels in some areas of the country, the primary source of the chemical is through active fluoridation campaigns. Surveys comparing cancer deaths in large U.S. cities where fluoridation started before 1960 with those in fluoridation-free cities suggest that the forced medication of the inhabitants results in 40,000 excess cancer deaths per year. Cities currently fluoridating their water include Chicago, Philadelphia, Baltimore, Cleveland, Washington, D.C., Milwaukee, St. Louis, and San Francisco.

Mercury is a lethal and ubiquitous form of pollution. It enters our waters as manufacturing wastes from things like paint, paper, vinyl chloride, and fungicides. Mercury's physiological effects include damage to the central nervous system and the kidneys.

Alaskan seals have been found to be carrying 116 times the amount of mercury considered safe for human consumption. The origin of the metal was traced to industrial wastes dumped into coastal waters off Oregon and Washington. The mercury moved up the food chain from plankton to salmon to seal, traveling some 2,000 miles before it finally reached the seals. This cumulative effect has poisoned many of the fish in the oceans, rivers, and lakes around the country and made them unfit to eat. Even fish like swordfish and tuna that have long withstood the effects of pollution are now showing levels of mercury contamination.

During the 1950's, a factory in Minamata Bay in Japan dumped great quantities of mercury salts into the bay. More than 100 people were poisoned. Forty-three died. Others went blind, suffered brain damage, or lost the use of their limbs.

For years it was assumed that mercury would never be a pollution problem because it was too expensive for industry to waste. We now see more than 6 million pounds of mercury used annually, almost half of which ends up in our waters. Even if we do stop mercury dumping today, without concerted efforts to remove existing deposits from our environment, it could be thousands of years before people could safely eat fish from a contaminated region.

Nitrate, a chemical by-product of nitrogen fertilizers commonly used on farms, is believed to contaminate about 20 percent of the nation's wells. In Kansas, 29 percent of farm wells sampled in 1986 contained high levels of nitrates, while drinking water in 50 percent of the communities showed levels of the chemical in excess of federal standards. Nitrates are associated with "blue baby syndrome," a potentially fatal blood disease in infants.

Polychlorinated biphenyls or PCBs: These chemicals belong to a family of substances called halogenated aromatic hydrocarbon. This family also contains such chemicals as DDT and TCDD (or dioxin), which are among some of the most toxic chemicals known to man. Like the chlorofluorocarbons (or CFCs) that are so damaging to the ozone layer of our atmosphere, the properties of these chlorine derivatives making them so attractive to industry are the same properties causing environmental havoc. They are extremely stable, so they remain in the environment for a long time, are heat-resistant, durable, nonflammable, and nonconducting. PCBs have been widely used in transformers and other electrical equipment, pesticides, heat-exchanger fluids, paints, plastics, adhesives, and sealers. High levels of these chemicals have entered waterways throughout the country either by industrial dumping or sewage systems. Some of the most contaminated areas include the Great Lakes; Escambia Bay, Florida; Waukegan Harbor, Illinois; the Chesapeake Bay; San Francisco Bay; Puget Sound; and the Hudson River.

Because these chemicals are heat-resistant and do not break down easily, they present many disposal problems. Today, the EPA has approved burning them in giant incinerators, but even at temperatures of 1000 degrees Celsius scientists have discovered that the temperatures merely drove the PCBs out of the contaminated sludge and into the gas vapor exiting the incinerators. Incomplete burning of PCBs is also one way in which dioxin is created.

Burying the chemicals in landfills or dumping them in water-ways, which served as the primary method of disposal for many years, has lead to widespread contamination. The EPA estimates that 90 percent of the world's population have measurable levels of PCBs in their bodies. Like DDT, these chemicals accumulate in the oils and fats of animals and increase in concentration as they move up the food chain. Hence, if the cycle started with tiny levels of PCB contamination in plankton, by the time you eat a fish that has eaten another fish that has eaten the plankton, you will be receiving a much more concentrated dosage of the toxic chemical. According to Joseph Highland of the Environmental Defense Fund, the levels of contamination and the number of people affected continue to increase every year. Human breast milk is currently so heavily contaminated that the average nursing infant exceeds by 10 times the maximum daily intake level of PCBs established by the Food and Drug Administration. Fish, birds, and livestock in many parts of the United States are literally sodden with PCBs.

Despite its danger to human health, PCBs are not on the EPA's list of thirty regulated chemicals. The agency does regulate the use and disposal of PCBs, but these regulations are so deficient that they have already been challenged once in court by the Environmental Defense Fund. The environmental group won the case, and the 1979 regulations were declared invalid. In 1982, the agency issued new regulations, which still allowed for the use of PCBs in existing transformers and called for a 10-year phasing out of their use in capacitators.

Dioxin is one of the most toxic chemicals known to man. We have already discussed it in some detail earlier in this chapter concerning paper-mill emissions and the production of dioxin as a by-product of chlorine bleaching of paper. This substance is so lethal that Canada has set its safe limit at 20 parts per *trillion,* while New York State sets an even more cautious standard of 10 parts per trillion. Despite the concern over even traces of this substance, the EPA still has not included it on its list of regulated chemicals. In concentrations as low as one part per billion, dioxin can be fatal; at lower levels (measured in parts per trillion or quadrillion) it can cause cancer, serious skin rashes, and a host of systemic disorders.

Citizens around the Great Lakes have been waging an ongoing battle with industry, particularly the giant Dow Chemical, concerning the emission of toxic substances that can combine to form dioxin. At Dow's Midland, Michigan, production facility, the company uses as much as sixty-three million gallons of water each day for its manufacturing processes. Dioxin is formed as a by-product of a number of chemicals, many of which are manufactured by Dow at this site and released in its waste water into small receiving streams. The company has even admitted that dioxin is present in the dust and soil surrounding its Midland site. Because of its extreme toxicity, even state-of-the-art equipment may fail to detect traces of the chemical. So, while licensing authorities regulating waste emissions have imposed stricter guidelines on Dow, without an outright ban on all dioxin-producing emissions, keeping levels of the chemical within safe limits appears improbable.

Chlorine was introduced in 1913 to disinfect water by killing bacteria and viruses. In a never-ending battle to fight pollution, water authorities across the country have been adding increasing amounts of chlorine to "purify" our drinking water. In Cincinnati, for example, by 1970, the Cincinnati Water Works had increased its use of chlorine 200 percent during a 15-year period. This may seem like a responsible reaction. If the water is dirtier, add more of a longtime dependable cleanser. Unfortunately, it does not work that way. Though a glass of water may look and smell clean as a result of chlorine treatment, it is becoming increasingly apparent that water is not made safe by remedial purification. Instead, the only way to have clean water is to keep it from getting polluted in the first place.

But that is only part of the chlorine problem. When it is used in high levels, chlorine can cause genetic damage and several forms of cancer. This is partly because of the toxic character of the chlorine itself, but also stems from extremely dangerous by-products that are formed when chlorine-treated surface waters interact with organic matter. These by-products, called trihalomethanes (THMs), are much more toxic than the chlorine itself and are recognized carcinogens.

According to the U.S. Council on Environmental Quality, "The wide practice of chlorinating public drinking water appeared to increase the risk of gastrointestinal cancer over an individual's lifetime by fifty to one hundred percent."

The number of chemicals threatening the safety of our drinking water are almost countless. I have attempted to familiarize you with a few of them and to give you some idea of how they can find their way into your water supply. These examples are merely illustrative. Drinking-water contamination can take place almost anywhere in the country. In Florida, for instance, where 92 percent of the residents rely on groundwater for drinking, state officials and environmentalists are now worried about the increasing number of groundwater sites discovered to be contaminated with highly toxic pesticides used by many citrus growers.

Temik, manufactured by Union Carbide, was long considered a dream pesticide by farmers because they could simply bury the granular substance at the base of a tree twice a year. Initially, Union Carbide claimed that the chemical would not filter down into groundwater. Later, however, scientists discovered that, absorbed by the plant's roots, Temik also seeps into the groundwater, where it can remain active for 5 to 20 years.

The groundwater around El Paso, Texas, is now so tainted with chemicals and salt that many people, unable to find potable water by digging wells, are forced to haul all the water they use from metered standpipes, usually located at some distance from their homes.

In May 1988, *The New York Times* carried an article entitled "Puzzling Findings on Bottled Water in Pregnancy." In that article, the *Times* reported that California health officials were expected to conduct further studies to "explain puzzling results that seem to suggest that tap water might contribute to miscarriages and birth defects." The initial studies, including 5,000 pregnant women and later confirmed by three other studies, found that women drinking bottled water during pregnancy had fewer miscarriages and a lower percentage of birth defects. Given what we now know about the safety of our water supplies, the real enigma in these findings is why scientists and reporters should be so puzzled by them.

Development, Misuse, and the Politics of Water

Up until recently, water has been largely plentiful. There has been no real need to worry about its future or its conservation. Today while contamination is threatening much of the nation's drinking water, the remaining potable water is being overtaxed by the complex interaction of a number of sociological factors.

Figuring prominently among these factors is the uncontrolled and ill-planned development now taking place throughout the country. In Houston, Texas, major sections of the city and surrounding areas have sunk nearly four feet, and geologists predict that within the next 20 years these city areas may fall an additional 8 to 10 feet. Whole areas surrounding Galveston Bay are now suburbs underwater, permanently lost. The cause of these sinkings are not earthquakes, floodings, or other natural phenomena, but rather the result of the overdemand on limited water tables due to rapid population growth and ill-planned development. Prior to the Houston experience, these sorts of cave-ins used to be unique to desert areas such as Phoenix, where it was easy to consume more water than could naturally be returned to the ground. This is due to what is called "mining" or "overdrafting" the water.

Fred Powledge, a journalist specializing in water problems, believes wholesale development of the countryside, or urban sprawl, has drastic effects on the water-system ecology, which, it should be emphasized, is the most fragile component of the ecological system. The sinking of whole buildings and areas into these giant potholes is due to the depletion of the water table, largely because of the excess demand placed on a limited water supply. When we pave large surfaces to build shopping centers, schools, and hospitals and encourage people to move there, we eliminate land that would otherwise effectively act as a sponge, soaking up rainwater and depositing it into the aquifers. The new population compounds this damage by increasing the consumption of an already decreased water supply. When these areas are near rivers, the problem is even worse. The paved areas act as a ditch leading directly into the river and causes flooding.

As supplies become scarce, water is becoming an increasingly political commodity. The growing competition for water is exemplified in Denver, where many competing interests are battling for control of the underground aquifer system. Wars are being waged by all sectors of society—livestock farmers, agricultural farmers, industry, builders (who without water cannot supply the needs of real estate developers), and municipalities fighting to maintain adequate supplies to serve the existing populations and structures. By the late 1960s and early 1970s, water had become such a coveted commodity that lawyers had already begun to specialize in and litigate over water rights around the Denver area. According to Powledge, even then competition for water was an issue. Subsidized agriculture and the associated irrigation contribute much to the water shortage and competition.

Ironically, much of this competition also springs from abundance. In 1902, the Reclamation Act was passed, paving the way for the massive governmental works projects of the 1930s that dammed the Colorado and Rio Grande Rivers, as well as many others throughout the country. It was at the time a grand project conceived during the Dust Bowl drought. By the mid-1930s, the Hoover, Shasta, Grand Coulee, Bonneville, and Fort Peck dams were simultaneously under construction. At that time, they were the five largest structures on the face of the earth. Over the following 40 years, nearly 1,000 other dams would be built.

These public-work projects did put America back to work and provided large quantities of water for agriculture in areas that could never have supported crops without it. But these massive undertakings also grossly intervened in ecostructures of entire states and started us on a road of waste and mismanagement that is now here to plague us. Mark Reisner, a former staff writer for the National Resources Defense Council and the author of *Cadillac Desert*, a history of water and the American West, points out some of the almost absurd situations into which our water policies have led us. In California, the single biggest consumer of water is irrigated pasture: grass grown in a near-desert climate for cows. In 1986, irrigated pasture used about 5.3 million acre-feet of water— as much as all 27 million people in the state consumed, including for swimming pools and lawns. Its financial contribution, on the other hand, was only one five-thousandth of California's $500 billion economy, while it drank up one seventh of the water.

Reisner points out that California is unusual only in that it uses proportionately *less* water than most other states serviced by the dam networks. In Colorado, alfalfa to feed cows consumes nearly 30 percent of the state's water, much more than the share taken by Denver, yet adds less than $200 million to the economy. Tourism uses almost no water but contributes about 12 times as much to the economy as alfalfa.

The politics surrounding water rights and usage are indeed complex. Once farmers become used to and dependent upon a cheap and virtually limitless supply of water for irrigation, there is little incentive to change. When there was enough to go around without conservation and the nation needed the jobs generated by the construction from these systems, the benefits appeared to outweigh the costs. But things have changed since the 1930s. Water is now in demand, and people are starting to look at what it costs us to maintain programs like irrigated pasture and rice growing in the desert.

There are a number of viable and affordable measures that could be taken immediately. Simply lining the earthen canals of the Imperial Valley with cement, for instance, could save nearly half the amount of water being diverted to Arizona. Southern California's Water Authority has proposed to assume the cost for this lining in exchange to the rights to all water saved.

As it stands now, farmers pay as little as three cents per 1,000 gallons of water. If a farmer installs water-saving equipment and decreases water consumption, or substitutes a less water-hungry crop, his neighbor can merely take what he does not use. Over time, if consumption stays low, that farmer's water rights will decrease proportionately—hardly an incentive to conserve. Under the free-market system, farmers would be allotted a certain amount of water credits, and they could sell what they do not use. That would result in a price stabilization at about 10 cents per 1,000 gallons. This would roughly triple the farmer's cost, but less than halve the amount the Metropolitan Water District would have to pay for lining the All-American Canal linking the Colorado River with the Imperial Valley. On the other hand, a free market in water rights would not only encourage conservation, it would also cause farmers to switch to the growing of crops more suitable to their areas. In the long run, taxpayers would save money. They might pay more for a head of lettuce or a steak, but they would also see the phasing out of a wasteful and irrational use of their tax dollars.

Solutions and Resources

The dichotomy existing with the Indian treatment of the Ganges is seen throughout the world. Rather than being beyond such practices, if anything, the industrialized world is even guiltier than the Third World in terms of wholesale dumping of waste into its waterways. We have the knowledge and the tools at our disposal to manage our waste in ways that will not pollute and defile our waters, but for the most part, we are either too lazy, too short-sighted, or too cheap to employ them. Surprisingly, cleanup measures do not have to be complicated or even expensive. More than 50 years ago, West Germans living in the Emscher Valley of the Ruhr River organized the *Genossenschaften*, an association dedicated to controlling pollution in the area. The move came in response to an epidemic of illnesses following the failure of municipal sewage-treatment plants. Today, the *Genossenschaften* has organized and oversees a complex project of pollution control. The guiding principle of the project is one of personal responsibility— those who produce the pollution accept that they must be responsible for cleaning up their own waste. Effluents are monitored and tested in order to determine the charges payable by each industry or municipality in the area. The *Genossenschaften* receives a steady income from the effluent charges and from the sale of drinking water. The income is used for the management of treatment plants, dams, storage lakes, aeration equipment, and land drains.

The effluent charges have advantages. Most important, they provide the funds required to do the cleanup in the first place. Many regions in the United States would like to clean up their waterways and beaches, but exorbitant costs are always cited as a reason for putting off the cleanup activities. When each user has to pay as it pollutes, the funds are available immediately and directly, as opposed to the monies coming out of tax dollars that are always subject to a large number of competing demands. Moreover, effluent charges are more equitable than making the general taxpayer support the cost of cleanup activities. The polluting industry or the city will surely pass the costs on to the consumer—either in end costs or things like sewage-related taxes—but there will also be more incentive to shift away from polluting activities. If, for instance, a company producing white paper and emitting dioxin as a by-product is forced to pay for those emissions, the cost of its paper will go up. But, if the paper is more expensive, people may shift either to paper that is not quite as white or one that has been bleached through a less environmentally damaging process. The West German experience showed that by having industries pay according to their amounts of effluents, many companies were motivated to develop technology to cut down their emissions of toxic substances. Many companies began to recycle wastes and found that they could make a profit in the meantime. For example, usable sulfuric acid is being recovered in some of West Germany's steel factories. Canning plants have found that they can recover salable vinegar from what was once waste. Paper industries have cut down their effluents ninety percent simply by switching from one process to another. In Japan and the Soviet Union, environmentally conscious managers have even found that thermal pollution can be put to good uses. They have developed systems where the hot water is pumped for cooling, for recreational purposes, and for fattening up eels and carp, both of which respond well to the warmth.

As simple and reasonable as the West German solution may seem, the adoption of any solution to water pollution is still the exception rather than the rule in this country. For the most part, progress is a result of citizen action rather than government leadership.

An enlightened public, armed with a positive can-do attitude toward water pollution, is not a substitute for strong political leadership, but, for the moment, it is the most effective means for getting things done. This goes contrary to what most people believe, namely that environmental issues are so vast and complicated that individual citizens cannot do anything about them. But we have seen time and time again that it is ordinary citizens uniting with others like themselves who are making the real changes in our environment. J. Larry Brown, director of the Community Health Improvement Program (CHIP) at Harvard School of Public Health, and Deborah Allen, senior program coordinator of CHIP, believe that in most cases, alarmed local citizens do what they are supposed to do; they go to local officials to express their concerns and request assistance. When they don't get it, the initial shock of the existence of toxins takes a backseat to the outrage people experience as their officials do nothing. In a classic form of beheading the messenger, town officials may charge the citizens with being radicals, with being insensitive to the economic repercussions of the issue, or even with seeking to foster fear and turmoil in their community.

Brown and Allen point out that it is usually at this point that the town splits into opposing factions—the concerned citizens and those who feel that the problem either does not exist or is overstated. Often this latter group is composed of city officials, workers employed by the polluting industry, and the industry threatening to shut down if an issue is made of pollution. The troublemakers are also often told that they have no proof that the toxins are causing the harm they claim. Then the people who first discovered the problem now bear the burden of proving it is a problem.

After having consulted the scientists and the experts and the politicians, citizens begin to understand that they don't have a scientific or technical problem, but a political one.

Below we will examine three case studies of towns or areas in which water-pollution problems have been discovered. In the first case of Acton, Massachusetts, you will see the concrete example of what Brown and Allen were describing—citizens discover the problem, and officials either fail to act or side with the polluting industry. The second case, that of Waukegan Harbor in Lake Michigan, shows the kinds of results we could be seeing in water-pollution issues if the government were diligent in its enforcement and prosecution of the laws. Finally, we will look at the town of Barnstable on Cape Cod, and how it has become a model for groundwater preservation.

Case 1: *Acton, Massachusetts*

In 1954, the W. R. Grace company opened a chemical plant in Acton, about 30 miles northwest of Boston. By 1973, complaints about chemical fumes and groundwater contamination were becoming fairly frequent. The company was found to be emitting formaldehyde vapors and, in 1981, a large release of styrene gas necessitated evacuation of local residents. While the EPA contended that air pollution was the company's major problem, it soon became apparent that its pollution of groundwater was at least as serious, if not more so.

In 1978, Grace applied for a license to open a new battery-separator plant at a location that was not far from wells supplying 40 percent of the town's water. Townspeople were opposed to the new construction. The Massachusetts Department of Environmental Quality Engineering (DEQE) took samples from the well that were about half a mile away from Grace's existing plants. The samples revealed levels of nine toxic chemicals, including benzene, toluene, chloroform, and methylene chloride. Results of the study were known to town officials, but they issued the permit to Grace anyway. Residents were not made aware of the study until two weeks later, when the wells were shut down.

Shortly thereafter, the Acton Citizens for Environmental Safety (ACES) was formed to protest Grace's waste disposal practices, which it believed to be responsible for the wells' pollution. The group was criticized from the outset and was told that it was overreacting. Although the wells had been closed for safety reasons, town officials said that the water was not *that* polluted and are on record as even suggesting that the chemical "may even be good for you."

With little public or governmental support, ACES found it was fighting an uphill battle. Grace's practice of dumping its load of toxic-waste water into unlined pits continued unabated. By the end of 1979, a report on the contamination of the well confirmed the citizens' concerns about Grace's practices. The water was poisoned from chemicals emanating from the Grace facilities. Public support began to grow for ACES, but government officials maintained that there was nothing they could do to stop the company from conducting its business as usual. Of course, this was untrue. Even at that time, there were laws that would have enabled environmental officials to order Grace to stop polluting, had the officials chosen to enforce them.

The events surrounding Grace and its pollution continued to be a major boondoggle. Two years after the well-water contamination was discovered and a year after Grace was cited as the culprit, the EPA and DEQE did manage to convince Grace to stop its disposal practices. But little was done to force Grace to assume responsibility for the mess it had already made. Instead of taking charge of the situation, the government allowed Grace to hire a consulting firm to assess the degree of pollution existing at its sites. Not surprisingly, the results of the assessment were as favorable as they could be to Grace. Nevertheless, some pollution was so pronounced that even a self-reviewing study could not hope to totally camouflage it. Even then, government dragged its feet. While dangerous metals like arsenic, beryllium, and chromium were found in high concentrations in the unlined lagoons, well waters remained untested for these substances. The Grace disposal sites in Acton were eventually designated a Superfund site. In 1987, the company agreed to pay Acton $2.35 million for past and future water treatment.

Case 2: Waukegan Harbor, Michigan

There is some progress being made in cleaning up our waters, and although citizen-action groups and grass-roots organizations are responsible for the lion's share of the credit, government agencies do make some progress when they are diligent in their enforcement of the law. In October 1988, after many years of bitter battle, federal and state officials succeeded in getting an outboard-motor manufacturer on Lake Michigan to agree to pay $20 million toward the clean up of PCBs it had been dumping into the Waukegan Harbor since 1961. Shortly following the signing of the consent decree between Outboard Marine Corporation and environmental officials, Illinois State attorney general Neil Hartigan hailed the agreement as a "major environmental victory," especially, as he pointed out, because the cleanup was being accomplished without the use of taxpayer dollars.

This case also points out the importance of government involvement. Citizens could perhaps have achieved the same results, but it would have been an even longer and more harrowing battle without the clout of the federal government. One of the reasons that the case dragged on for such a long time was the company's obstructive tactics. According to EPA spokesperson John Perrecone, the company first claimed that PCBs presented no health hazards, then later, when federal officials were seeking to determine the extent of the pollution, the company denied the officials access to the plant. In 1986, however, armed with an amended federal waste-cleanup law allowing them access to the premises, the federal environmental officials did manage to get on-site. When the company could no longer hide behind trade secrets and the privacy of its property, it realized that it had no real alternative but to negotiate.

Case 3: Barnstable, Massachusetts

The town of Barnstable, Massachusetts, on Cape Cod provides another good example of how much can be accomplished when all levels of government cooperate in issues affecting our waters. Barnstable, once a sleepy little beach community, has over the last 20 to 30 years experienced one of the fastest growth rates in the country. Between 1960 and 1980, the town saw a 129 percent increase in its permanent residents, and a 185 percent growth in summertime inhabitants. As with all areas of the country undergoing this rapid growth, one of the major problems facing the community has been to ensure an adequate supply of safe drinking water. While Cape Cod has a natural abundance of high-quality water, its sandy soil makes the underground aquifers particularly susceptible to pollution by contaminants. By 1988, town officials realized that they were facing a serious water-supply crisis—six toxic dumps had already been identified as sources of pollution in nine of the town's public water wells. Instead of procrastinating until the severity of the problem made it virtually unsolvable, the town took immediate action. It bought up over 774 acres of endangered land for some $32 million, ordinances were passed to place tighter controls on polluting activities, and a public awareness campaign was initiated. The town also took its case before state and federal officials. With support and cooperation from all levels of government, Barnstable initiated a comprehensive program of groundwater protection that may serve as a model for many other areas of the country in years to come.

The Cape Cod Aquifer Management Project (CCAMP), an association of water officials from communities all along Cape Cod, supervised the project and developed a workable and realistic approach to the problem. It first looked at the nature of the resource—its vulnerability to contamination, the demands placed upon it by population growth—as well as mapping out specific areas that supplied well water for the town. It then looked at land use in the area. Scientists were called in to examine existing sites believed to be contributing to pollution. One specific site was chosen for an in-depth analysis. Although it was around a major wellhead providing much of the town's drinking water, the CCAMP found 186 underground storage tanks, 6 confirmed hazardous-waste sites, and a waste-water-treatment plant that had significant runoff into the ground. Armed with information like this, all levels of government began to work together to find solutions. City planners participated in drawing up zoning regulations that would minimize further pollution, law-enforcement officials were given the go-ahead for a clampdown on polluters, and standards were set for the disposal and storage of toxic substances.

Cleaning up our waterways and oceans is going to necessarily entail the development of adequate waste-disposal systems. The first and most important step here is to control pollution at its source. As we have seen, West Germany has found a "polluter" tax as one viable way to achieve this end. With polluters each assuming responsibility for their own waste, with time they inevitably find ways to decrease emission. Going hand-in-hand with this approach is strict government enforcement of pollution laws. We have the laws to require polluters to stop pollution and to pay for what they have already done. It is up to each of us to let our politicians know that we expect these laws to be enforced and that we do not find it acceptable to continue picking up the bill for corporate America's irresponsibility. Just following the Alaskan oil spill, Exxon announced that it was assuming full responsibility for the disaster. A few weeks later, stories began to surface that Exxon would avail itself of a tax write-off for cleanup costs. This is currently legal under existing laws, and Exxon is not the first, nor will it be the last, to use such laws. While Exxon could have gone a long way in improving its public image by choosing not to use these business write-offs, it still was not violating any law in doing so. If you feel that this sort of double taxation on the public—first the loss of the natural resource, then having to pay for it—is not an acceptable way to have business conducted in this country, it is time to speak up. You can do this either through direct communication of your feelings to your political representatives or by joining any of the numerous environmental groups who feel the same way you do.

For people concerned with the quality of their drinking water, there are a number of high-quality water filters and purifiers now out on the market. We suggest, however, that before buying one, you look at the literature and do some research to make sure that you are cleaning rather than contaminating your water as you filter it. In September 1988, the Federal Trade Commission (FTC) issued a ruling finding that Norelco had shown "utter disregard for the welfare of its consumers and blatant disregard for the law" in knowingly marketing a water filter containing a recognized carcinogen. The Norelco fiasco illustrates not only gross corporate irresponsibility, but also the failure of governmental agencies to prevent such activities. Between 1982 and 1986, the company sold 186,000 filters for its Clean Water Machine with the knowledge that the filters could contaminate water with the carcinogen methylene chloride. Four government offices have overlapping jurisdiction over methylene chloride and none of them did anything about its presence in the Clean Water Machine. Finally, three years after Norelco stopped manufacturing and advertising the water purifier, the FTC filed a suit last year challenging the company's advertising claims.

This is not meant to discourage you from buying water purifiers, only to suggest that you might want to do some investigation before spending your money on a particular unit. There are many on the market, and most are reputable. The Multi-Pure system sold in many health-food stores is a
particularly fine product for chlorine and bacterial removal.

The same caution should be exercised when buying bottled water. New York State has recently upgraded its laws for testing bottled water. Many states require water bottlers to be certified in order to make claims as to the water's purity. Often these certifications will take the form of a number or a code on the label of the bottle. Most reputable companies do regular quality checks on their water. Deer Park Spring, a company distributing along the East Coast, says that it tests its water every four hours for contaminants. Evian, the world's largest bottled water distributor, says that it does 200 checks a day on its water.

If you are concerned about the water in your area, there are a number of environmental organizations throughout the country that may be able to advise you on your particular problem.

• The Environmental Policy Institute, (202) 544-2600, is located in Washington, D.C. At the time of this writing, the person to contact concerning EPA landfill regulations and groundwater pollution is Valma Smith. She is currently involved in trying to enact a national law to protect groundwater.

• Another Washington-based organization is the Clean Water Action Project, (202) 547-1196. Eric Johnson, a member of the or ganization, has a particularly good understanding of water issues facing the country.

•Public Citizen's Group, a Ralph Nader organization, (202) 546-4996, also works via its lobbyist, Anne Bloom, toward passage of a uniform ground-water protection act. She specializes in ground-water pollution from things like pesticides, landfill, and solid-waste leaching.

Part Two
The Food We Eat

Chapter 5
Modern Farming

You don't have to be fiercely nationalistic to look in wonder upon the great riches and natural wealth of this country. This is particularly true in terms of food-producing resources. America has some of the most fertile and productive farmlands in the world. Even before the advent of the enormous agribusiness industry, we were not only self-sufficient, but also a great exporter of food. But today there is another side to the food story, involving some of the largest and most influential sectors of society, ranging from the federal government to General Foods and Coca -Cola; from petrochemical producers and biotech firms to farmers and ranchers. This story includes the farm crisis, and the eroding and decreasing quality of our soil, as well as the pollution of our lakes and streams. Unlikely as it may seem, the way that America eats and produces food is even related to the rampant deforestation now taking place in many parts of South America.

While this country does export nourishment in the form of grains, its major contributions to the world food supply today come from the export of much more profitable products such as pesticides, fast food, and soft drinks. Accordingly, the history books of the future may not see America so much as the world's breadbasket, but rather as a major exporter of poison, tooth decay, heart disease, cancer, malnutrition, and a plethora of other health and environmental problems now linked with America's chemically produced high-fat high-sugar and high-protein diets.

Most Americans cannot even conceive of what it means to be without an abundance of healthful and varied foods. We walk through our supermarkets, and the shelves are always full. Driving across country, food-growing areas extend as far as the eye can see. It was precisely these vast expanses of rich and fertile farmland that attracted many of the early settlers. But most people are completely unaware that America's cheap and abundant food supply hangs in a precarious balance, threatened by factors so intricate, it is tempting to shrug one's shoulders and walk away from it all. But can we afford to do that? The answer is "obviously not," and fortunately there are a number of things that can be done to reverse the trends.

Portraits of Farming in America

Increasing pressures of all kinds in the Farm Belt have given rise to widespread despair. A new movement of mental-health care has been created, devoted to the growing number of farmers who battle severe depression and suicidal urges.

According to rural sociologists, an increasing number of farmers are choosing suicide rather than dealing with the growing hardships facing them. A survey released by the Agriculture Department in 1985 showed 31 percent of American farmers to be "economically threatened," with less than enough income to pay their bills. These survey results were issued even before the farm crisis of 1986.

In California, which boasts a nearly yearlong growing season that can support more than 200 crops, farmers have not escaped the financial choke hold on American agriculture. Falling commodity prices, fewer exports, and a sharp drop in land values are taking a toll despite ideal growing conditions. In 1986, California real estate agents noted that there had not been such a glut of farmland on the market since the Depression, as farmers falling victim to hard times began unloading parcels of land in an attempt to meet their loan payments.

Many farmers are able to make more money selling their land for *nonfarming* uses than by struggling to grow crops for which there may be no market. Developers already buy 12 square miles of prime farmland every day, paving over more than 3 million acres of productive soil yearly to accommodate advancing urban sprawl. In California, where land speculation is second only to that of New York City, bankers and investors see prime farmland as the opportunity to make a fast buck. A grape grower in the Fresno area who had farmed the land his parents moved to in the 1920s was told by his loan officer that he would have to sign over his farm to the bank when he missed one payment on his bank loan. Perhaps as many as 20,000 farmers face being thrown off their land in just one year. Half the farmers in the state—the ones who account for the bulk of production—could be replaced by banks and other financial institutions. As this occurs, farmers with a history of agriculture in their blood and a love of the land are being replaced by absentee owners who hire managers to exact maximum profits from the land.

Cornelia Flora, a professor of sociology at Kansas State University, equated the conditions of many Kansas farmers with those of the Third World. "The malnutrition and hunger we're seeing occur because people cannot earn a living in their own towns, and they are too poor to go to the cities," she says. In Nebraska, doctors are seeing diagnostic evidence of malnutrition in farm children.

These crises also came before the summer drought of 1988, which dealt the final blow to the many who were already economically strapped. How can a crisis of this magnitude be affecting American agriculture, long considered an eighth wonder of the world? What does it mean to America as a nation to see its farmers— symbols of integrity, hard work, and stability—suddenly teetering on the brink of extinction? And how can we seemingly ignore such a threat, not only to the individual farmers, but to the integrity of our entire farming system?

Revolutions in Farming—Agribusiness and "Farms as Food Factories"

Nowhere is the dichotomy between what modern agriculture believes itself to be and what it actually is clearer than in the contrast between the expected and the actual results of farming in America today. Despite the difficulties faced by America's farmers, the country has experienced a tremendous surge in food productivity. Much of this can be traced to "the green revolution" that took place after World War II and dramatically altered the face of American agriculture. With this revolution came the industrialization of farming and virtually all aspects of food production. Its ultra-modern machinery, hybrid seeds, fertilizers, pesticides, and sophisticated irrigation systems permitted the productivity of land to triple between 1950 and 1970.

Farming became "agribusiness" as large companies, capitalizing on bumper harvests that gave high returns on investment, began taking control over U.S. food production. At first, this made good sense. American business had long prospered by taking advantage of what the economists call "economies of scale," meaning that the more a company produces of a commodity, the lower the cost of production of each additional unit. Industries like oil and steel had done very well by taking advantage of the profits realized through size. So why should it be any different with food?

For all but the most astute observers, high yields and abundant crops did make it seem that food could be produced just like any other commodity and that the American farm could be operated like any other business venture. American farmers were told by Earl Butz, secretary of agriculture during the Nixon administration, to "get big or get out." The federal government was one of the major supporters of large-scale, centralized agriculture, encouraging the trend toward bigness through subsidies, tax credits, and research grants. Top agricultural officials, one after another, issued repeated praise for the concept of the farm as factory. Orville Freeman, secretary of agriculture under the Kennedy administration, hailed American farming as "the greatest production plant in the world." One former Assistant Secretary of Agriculture propounded the idea of agripower. "True agripower is the capacity of less than five percent of America's population to feed itself and the remaining ninety-five percent with enough food left over to meet market demands of other nations," he said. "The real measure of agricultural strength is productivity combined with processing and marketing efficiency."

The price of land soared twelvefold between 1950 and 1979 as pesticides, mechanization, and single-crop plantings increased per 1-acre production, and profits. The number of farms grossing more than $100,000 annually increased 20 times, but in the meantime the total number of farms decreased by half. In a recent report issued by the Office of Technology Assessment (OTA), predictions were made that by the year 2000, three-quarters of the nation's corn, wheat, cotton, wool, beef, chicken, tomatoes, lettuce, beans, apples, and a number of other staples will be produced on only 50,000 large farms. By comparison, in 1987 that same three quarters was produced by 650,000 farms and ranches.

The surviving giants will not be farms as much as factories employing the latest gene-inserting, embryo-transferring, electronic-monitoring, and chemical-dependent technologies. Meanwhile, in less than 15 years 1 million farms will have disappeared, according to the OTA. These will be mostly small- and moderate-sized operations that cannot afford to invest in the expensive emerging technologies outlined in the OTA report. That 1 million farms is more than 40 percent of the total number of farms and ranches in the nation today.

Here are some other startling facts and figures about food production in America:

• Approximately 7 percent of all farms control 56 percent of the nation's agricultural production.

• Fifteen agribusiness corporations provide 60 percent of all farm supplies.

• About 60 companies handle 70 percent of the nation's food processing.

• The USDA predicts that by the year 2000, 1 percent of all farms in the United States will account for 50 percent of all food production and that 4 percent of the farms will hold 60 percent of all farmland.

Rising Costs

According to a recent study conducted by economists at the USDA's Economic Research Service, in an attempt to maximize profits by higher yields, farmers are tending to use ever-increasing amounts of *off-the-farm inputs*, particularly pesticides and fertilizers.

The USDA economists estimate that in 1987 herbicides were used on more than 95 percent of corn and soybean crops and 60 percent of wheat in this country. Total pesticide use in 1986 reached an estimated 820 million pounds (of active ingredient), up from 335 million in 1965. Sales of pesticide chemical were estimated at $4.1 billion in 1987.

The aggregate use of nitrogen fertilizer has more than doubled since 1965—increasing from 4.6 million tons to 10.3 million tons in 1987. And many of these chemical fertilizers work like an addictive drug on the soil. They do their job; they give the earth the chemicals that plants need, but in so doing kill the natural nitrogen-producing bacteria. Fertilizer works terrifically on worn-out soil, but soon thereafter, nothing will grow without it. As Robert Thompson of the USDA recently put it: "The present system is simply pushing farmers to pour more and more costly inputs [fertilizers, hybrid seeds, herbicides, pesticides, etc.] into smaller and smaller levels of acreage. The result is that the actual costs per bushel are being driven way up, even as the values of land are being driven way down."

When one considers the energy gobbled up by modern agricultural practices, all this productivity does not appear so spectacular. Fertilizers, pesticides, tractor fuel, farm electricity,

food processing, and food transportation all require fossil fuels—
gasoline, oil, natural gas. For every person in the United States,
farm machinery, chemicals, and electricity used for farming consume the equivalent of 78 gallons of gasoline each year. Food processing uses another 37 gallons. Agriculture has come to consume
12 percent of this country's total energy supply, more than any
other single industry.

A careful examination of the *food factory* of modern agribusiness
reveals that as many as 10 calories of energy are required to produce each calorie of food. You do not have to be economically
astute to realize that this type of return on investment is a losing
proposition and cannot survive for long.

These increasing costs are creating a very peculiar situation. On
one hand, the nation is awash with food, but because it is so expensive, many U.S. food processors are going abroad for their produce. Outside the country, lower labor, land, transportation, and
other associated costs are permitting imported food to reach the
market at prices that American farmers cannot possibly beat. The
United States has become the world's second largest importer of
food (behind West Germany). According to retailers, 20 percent of
our fresh fruits and vegetables are imported, and many of them are
not easily identified as such. Many of the largest American food
companies, including Campbell Soup, Pillsbury, General Foods,
and Coca-Cola, rely heavily on imported ingredients. Canada,
Mexico, and a host of South American companies provide the bulk
of our food imports. Since 1980, imports have doubled, and experts believe that it will not be long before they outpace America's
food exports.

Supply and Demand in Topsy-Turvy
How can farmers be going bust when crops are booming? Ironically, in many ways, bigger harvests mean *more* problems for farmers, not less. While on one hand, farmers see their fixed costs rising,
what occurs when yields are particularly abundant is that the price
the farmer gets for his crop decreases— the basic law of supply and
demand. If economic analysis stopped there, the farmers could do
a number of common sense things to recuperate after a price fall.
They could, for instance, switch to growing another crop. Or they
could try to cut costs, as many farmers around the country are
doing, by decreasing their dependency on purchased inputs like
pesticides, chemical fertilizers, and heavy machinery. They could
also start to reexamine the fundamental precept of farming in

America today, namely that higher yields mean higher profits. Although some farmers have begun to do this, a fundamental switch in focus and attitude is necessary—a switch that agriculture officials are not willing to promote.

Senseless distribution of food resources affects the world. We now produce enough grain to provide every person in the world with two pounds each day. Yet, according to the World Bank, one third of the people in 87 Third World countries do not consume enough food to lead normal, productive lives. Nearly half of them, 340 million, are considered acutely malnourished and live on diets that stunt growth and severely threaten health. Between 1970 and 1980, the number of undernourished people in the Third World grew by 14 percent. "Increasing the food supply will not eliminate the problem unless it also improves the incomes and purchasing power of the poor," concludes a recent World Bank study.

Meanwhile, hunger is increasing throughout the United States. Researchers at Harvard have noted a return to levels present in the 1960s. According to the University-sponsored Task Force on Hunger in America, 20 million people go hungry each month in this country.

Industrial Stress and the Vulnerability of America's Food Supply

Modern agricultural practices, while yielding bumper crops under ideal circumstances, have also left American agriculture in an extremely vulnerable position.

American farming can be very productive through its dependence on strong plant species and extensive use of chemical fertilizers and pesticides, but 1988 demonstrated that modern industrial farming is also vulnerable to drought. With even hotter summers predicted for the future, huge crop losses could result. Food shortages could occur unless U.S. agriculture adapts itself to environmental change.

America's farmlands, like the rest of our environment, are also beginning to show signs of industrial stress. While the green revolution has undeniably brought about tremendous increases in yields, with time, each advance is creating its own set of problems.

Monoculture, the practice of planting large areas of land with a single crop, does increase per-acre productivity, but it also makes crops much more susceptible to insects. Consequently, farmers have needed to rely on ever-increasing amounts of pesticides to keep the pests in check. Once heralded as wonder chemicals, pesticides are now seen as at best a short-term solution to the pest problem. They are nonspecific and thus kill not only the target pest, but also the beneficial insects that prey on the pest. Ironically, in many cases, not only do the pesticides eradicate the natural predators, but the targeted pest also develops an immunity to the pesticide. Farmers' attempts to keep infestations under control by applying ever-increasing amounts of pesticides have little effect except to drain the farm's finances further and to pollute the environment. According to Robert Metcalf, a professor of biology and entomology at the University of Illinois, "Cotton bollworms weren't a big problem until farmers decided to spray them. Most of them were fairly well regulated by their natural predators, and when we started throwing pesticides all over the place, we killed our friends." Metcalf also points to the destructive Colorado potato beetle, which has been assailed with as many as 15 different chemicals since 1950 and now enjoys immunity to all of them.

This country once enjoyed some of the best topsoil in the world. This soil, developed over thousands of years as rocks and sediment are gradually ground into rich loam, is an invaluable natural resource. Instead of cultivating and nurturing the asset upon which their lives depend, our huge industrialized farms have in the space of some 30 years turned vast areas into wastelands. Large-scale irrigation projects, upon which much of today's agriculture relies, flood farmland with massive amounts of water, causing the buildup of salt and mineral deposits as the water evaporates, eventually rendering lands useless. Intensive planting also contributes to soil erosion.

It has been suggested that the use of hybrid seeds and monoculture works much like pesticides in rendering our agriculture ever more vulnerable. While hybrids have tremendously increased the tonnage of corn and wheat produced during the last fifty years, an unseen side effect is just beginning to surface. The systematic cultivation of only certain hybrids has eliminated the vast genetic variety that once existed. When a uniform, genetically identical crop is singled out for attack by a certain pest or disease, an entire crop may be ruined.

Just over 10 years ago, as much as 50 percent of the corn crop in parts of the South was wiped out by leaf blight. A subsequent report by the National Academy of Sciences said that the plants were "alike as identical twins."

The drought of 1988 drew attention to the need for more agricultural study. Researchers aim to broaden the variety of crops American farmers grow and widen their adaptability to more regions at a lower cost. This would protect American agriculture from harsh weather or other shocks. But our current farming practices, supported by the Agriculture Department, are what caused the decrease in seed diversity in the first place. Unless priorities are changed at the top level of agricultural policy-making, all the research in the world will not be able to stop the trend toward monoculture and the destruction of genetic diversity.

Genetic Engineering—Designer Animals and Miracle Seeds

Nowhere is the issue of genetic diversity of more important than in the up-and-coming area of biotechnology. Breeders have used a variety of techniques—and are now beginning to use genetic engineering—to produce crops that grow faster, are less expensive to plant, and have better defenses against insects, diseases, and harsh weather.

Dennis T. Avery, an agriculture analyst at the State Department, believes that the plant genetics revolution is affecting 90 percent of the world's land and 4.5 billion people. Seeds are better. They are easier to develop. And farmers have little trouble using them. In fact, the developing countries are getting more effect from this new revolution than the affluent ones.

In general, the press, government officials, and agribusiness are receiving the concept of biotech seeds and animals with great enthusiasm. In the *Yuba-Sutter Appeal-Democrat*, a local paper serving farming communities in the Sacramento Valley, two articles in one day's Farming section tell of the promises of genetic engineering. One article concerns the splicing of a bacteria gene to a tomato seed that would allow the seed to ward off tomato worms and other caterpillarlike pests without the need for pesticides. David Hulst, director of Hulst Research Farm Services, where field trials are being conducted on the seed, explains that "when the insect eats the plant, it ingests that bacterium, and dies." And he adds that the bacteria is only harmful to organisms with a specific alkaline level in their systems, which, he claims, will limit its danger to the caterpillar family. "It will not hurt any other living organism, only those that meet that criteria," says Hulst. Researchers for Monsanto Company, owner of the patent on the biotech seed, referred to the gene as a "naturally occurring bacteria," and suggested that "any vegetables that have caterpillar insect problems could benefit from this technology."

The other article in the California paper concerned a report by the Research Service of the USDA on what a future with biotech farm animals may look like. The following predictions, drawn from the report, illustrate the direction in which the federal government is encouraging agriculture and ranching to go:

•Steer headed for consumer markets could be spliced with a gene that would allow them to thrive on low-grade crop residues and fibrous plants that are barely digestible now. Scientists hope to find a microorganism that can break down lignin, which binds fibers like trees together. Or, says the report, scientists may be able to "borrow a gene from termites and insert it in a microbe that already lives in the ruman (stomach) of cattle."

•Disease is a big problem with our current methods of poultry- and livestock-raising because cramped and unsanitary conditions cause infection to spread quickly. Biotech chicken and cattle could be engineered with a hereditary gene that would make them resistant to disease—without poultry producers having to alter their practices in the least. Another possibility is the development of a chicken that could survive in a totally closed environment, thereby eliminating infectious diseases. This would also decrease the feed costs of each chicken, as food calories would not be wasted in keeping the bird warm.

•To increase production in the poultry sector, the report suggests the development of special breeder hens that could double their current rate of 140 chicks per year through artificial insemination. Also a possibility is to engineer a new dwarf breeder that could increase production while decreasing food costs.

The administrator of the USDA's research agency, R. Dean Plowman, warns us not to get our hopes up. "Of course, there are no guarantees that what the scientists envision will become reality," he says. "But based on the research they're involved in right now, it's certainly within the realm of possibility."

Genetic engineering is definitely on the rise, enjoying widespread acclaim and encountering little opposition. In July 1988, scientists injected 2,200 cornstalks with a gene-altered vaccine to test its efficacy against the destructive European corn borer. The experiment was the first approved by the EPA and USDA for outdoor testing of a genetically altered plant vaccine. Many are heralding the biotech solutions to pest problems as the new alternative to pesticides.

But are these new products really safe, or even safer? When pesticides were first marketed, the manufacturers assured us of their safety, explained that the chemicals would remain cooperatively at the base of the plant and would not even enter the soil, much less the crop itself or the groundwater. Despite all these guarantees, little more than 30 years after their introduction we are seeing them responsible for groundwater contamination, eutrophication of our lakes and streams, and causing cancer throughout the population, and moreover, they are not even effective at controlling the problem for which they were designed in the first place. Will genetic engineering follow the same route?

Viruses and bacteria are extremely dangerous organisms with which to be experimenting. AIDS and many other serious illness are related to microbial infection. Viruses are particularly adept at mutating, making the development of a vaccine to counter them impossible with the current state of science. Who is to say with certainty what the effects of a microbe-injected corn or bacteria-spliced tomatoes will be on human health? What if these benign microbes mutate? What if the alkaline content of a person's stomach becomes abnormally elevated at the same time that he eats bacterially treated tomatoes? These questions and more need substantially more research than they have received before the American public should accept claims of the safety of biotech seeds or genetically engineered agricultural practices.

If safety could be established, it is true that genetically engineered seeds could become valuable additions to agriculture, but only so long as they are used to supplement the world's seed bank, not to monopolize and eventually supplant it. An analogy can be made here with the practice of medicine in this country today and its reliance on pharmaceuticals. Synthetic drugs, antibiotics, and other patentable remedies all have their place in health care, provided they are used in moderation and, more important, when safer and nontoxic treatments have proven of no avail. Although I have long been a proponent of alternative courses of therapy, I certainly do not deny the value of things like surgery and antibiotics when they are absolutely necessary. But what we have seen in the practice of medicine today is an undue reliance on the use of synthetic drugs to the exclusion of other, more healthful healing modalities such as dietary changes, vitamin and mineral therapy, and cleansing regimens.

In the realm of farming, with the emergence of biotech seeds, we are looking at an even more macabre proposition—a situation where the alternatives may be eradicated at their source. Just as the pharmaceutical companies have made substantial inroads in decreasing the practice of a more holistic form of medicine, biotech-seed companies have enormous profits at stake in making their patented seed-supplant the use of all other varieties of seed. Re-member, biotech seeds, under current law, can be patented. Natu-rally occurring seeds cannot. That means that the company devel-oping the mutant seed owns it and can alone profit by its sale so long as the patent lasts. At first glance this may sound like no big deal. If the seed is better, then why should it not be used world-wide? But what is better? And whose definition of "better" is to be prevailing?

From an agribusiness point of view, better means greater yields, and better resistance to insects and climatic changes, but from a consumer's point of view, nutrition and affordability may be more important. While there has been plenty of hype about the new higher-yield seeds, little has yet been done to study how the nutri-tional value of crops fares when all emphasis is on profitability. Higher yields may mean for a time that food is less expensive, but what happens when all competition is usurped, and there are no other tomato or corn seeds but those held in patent by one com-pany?

Modern agriculture works well in ideal conditions, but often is extremely vulnerable if any of the variables that make up these conditions change. This is even truer when it comes to a biotech seed. The seed will flourish when the conditions for which it was developed prevail, but nature is not and never has been a constant. She has, however, provided for her inconstancy by creat-ing a diversity of species. Through diversity and adaptation, na-ture sustains life. By substituting man-made seeds for those devel-oped over thousands of years, we are essentially saying that we are willing to take the risk that man knows better than nature how to survive.

Because of agribusiness focusing only on high yields and other factors that bolster profitability, many of today's crops are sorely lacking in overall quality. While it is not impossible, for instance, to develop a high-yield seed that also also produces a flavorful and nutritious fruit, many of today's crops are measurably deficient in both—wheat has lower protein and vitamin content. Hybrid and biotech seeds are also developed to make foods more marketable. Fruits and vegetables must be able to travel long distances, sustain bruising, and still arrive at their destination with a just-picked, fresh, appetizing appearance. If you have bought things like super-market peaches or tomatoes lately, you undoubtedly have noticed that appearances can be very deceiving. Despite their fresh, whole-some colors, these fruits are often pithy, watery, and tasteless, bearing little resemblance to the fruits we once had.

Alternatives and Substitutes

American agriculture is teetering on the brink of disaster and getting more vulnerable every day. This is not meant to be an alarmist statement, but it is intended to get your attention. As the small- and medium-size American farmer becomes an anachronism, the changes affecting the American farm are also affecting each and every one of us. Food grown on factorylike acreage, harvested by huge machinery, sprayed, fumigated, irradiated, and then processed and packaged, has little in common with vine- or tree-ripened fruits, or organically grown and stone-ground grains. It becomes a dead and lifeless consumer product, not unlike any other. It can be traded and bartered on the stock exchanges, advertised on the television. It can be made to appear uniform, is easily packaged and sold as a bulk product, and in many cases has an indefinite shelf life. Breads like Wonder, for instance, have shelf lives of more than six months. In Europe, even day-old bread is looked upon with scorn. Perhaps worst of all, food ceases to have any connection with nature, losing much of its ability to heal and sustain us.

Despite the claims of agriculture officials, American agriculture has never been more vulnerable. What emerges from this picture is that American agriculture is not a sustainable, or even desirable, system. It is both polluting and exhausting to the environment. Our current agricultural system survives only because of government-support programs, which of course are supported with taxpayer dollars. If, for example, you add in the real environmental and social costs of irrigation to a head of lettuce, its price can range between $2 and $3—not the 69 cents marked in the supermarket. And that is only irrigation costs. This situation can only get worse as biotech companies develop seed monopolies, enabling them to set prices as they see fit.

As with so many of our environmental problems today, though the problems grow in magnitude, little help is forthcoming from government and corporate officials. This is no surprise, for they are the ones who created the system responsible for the problems in the first place. Fortunately, grass-roots movements are beginning to respond to these farming crises on their own. Their solutions are simple, and amazing because of their simplicity.

Growing attention is being focused on practices that will reverse and correct some of the problems created by conventional agriculture. Farmers are realizing that the key to their survival, and the survival of American agriculture as a whole, lies in decentralization and the development of sustainable farming techniques. In the Northeast, where the average farm is one third the size of the national average, many farmers have chosen to stay small, have begun to diversify their crops, and are selling their produce directly to the consumer at hundreds of roadside stands and farmers' markets. Compared with their midwestern peers, many New England farmers are prospering. The average acre farmed in Massachusetts, for example, produces nearly four times the amount of cash receipts that an acre in Iowa planted with corn for cattle does.

Some of the more creative and innovative midwestern farmers are beginning to take note. Their results are encouraging. In southern Minnesota, where 1,000-acre farms are going out of business all around him, one farmer and his family make a living from organically farming *five* acres. He now sells his own vine-ripened tomatoes, pumpkins, broccoli, cabbage, squashes, and melons to 24 local customers, including one wholesaler. He also grows greenhouse flowers, bedding plants, hanging baskets, and vegetable starts in a 30-by-50-foot greenhouse netting him more than $16,000 yearly.

Other farmers in his area have taken note. In order to avoid the competition that would occur if neighbors began competing for the same local markets, they've developed a system of selling larger quantities of pooled produce. A grocery-store chain in suburban Minneapolis has agreed to buy vegetables from the local farmers. Additionally, the government of Minnesota is lending them $3.5 million to build a food-processing center that will gather, prepare, and ship all the produce, and to construct a huge greenhouse where the farmers' plants can be started. The loan will also provide new growers with finances to start their first season at an enterprise that will turn a frustrating past into a fruitful future.

The Agriculture Department estimates that 30,000 of the nation's 2.1 million farmers are raising crops without chemicals. Other experts believe that the numbers may be as high as 50,000 to 100,000. Farming without chemicals, once considered a fringe movement, has gained ground and is moving closer to the mainstream of American agriculture.

The new techniques go under a variety of names—*regenerative, sustainable, low input, integrated pest management,* and *organic;* the latter being the least favored because of its negative connotations as a fanatic, extreme practice. Whatever the methods are called, their growing acceptance is encouraging. The USDA now offers information on sustainable agriculture through a toll-free number in Memphis. Programs have been established across the country to educate farmers about the new techniques. The state of Minnesota is showing particularly strong leadership in the area of sustainable agriculture. In June 1987, Governor Rudy Perpich signed into law a provision for the first university chair in sustainable agriculture at the University of Minnesota. Other universities are beginning to offer programs and research grants in the new techniques. In 1988, the Minnesota State Legislature approved landmark legislation proving a record $3.45 million in state-support funds for sustainable agriculture. According to Senator Patrick Leahy, who chairs the Senate Agriculture Committee, "This kind of research could eventually bring about the healthiest agriculture in the world."

Here are some examples of how a sustainable system can work: Strategically timed applications of manure can provide soil with almost all the nutrients it needs, thus limiting the need for chemical fertilizers. Rotation plantings will keep out weeds and bugs. After corn is harvested, a crop of oats which grow very thickly can be planted to crowd out and overpower weeds. Next comes alfalfa, which adds nitrogen to the soil, the active ingredient in fertilizers. With these practices, farmers estimate that they can produce a bushel of corn for 40 cents, one fifth the cost of growing the same amount with chemicals. The alfalfa crop averages six tons per acre, 50 percent better than state averages.

Forty miles away from the farms just described, an organic-vegetable and melon grower plants orange marigolds among his green peppers. The flowers act not only as a lure for pollinating, but also produce a natural chemical that repels pests. Through ingenuity, this farmer has discovered that the destructive potato beetle (the pest that has now developed immunity to most chemical insecticides) prefers eggplants to his potatoes. Mr. Ruehling's solution of interspersing his potato crop with eggplant is particularly significant because potatoes act as a sponge for all pesticides and agricultural chemicals. He has also found that onions and garlic act as a repellant for a variety of farm pests.

Following the summer drought of 1988, researchers in Salina, Kansas, began experimenting with techniques designed to duplicate conditions that allow grains to grow naturally on the Plains even in dry conditions. The idea is to reconstruct the prairie environment, which thrives in drought or wet weather, heat or cold. Prairie soil stores water. Conventional farm fields, which are subjected to mechanical tilling, pesticides, and fertilizers, cannot store water and often just wash or blow away.

Natural Food and Farming magazine recommends a number of innovative methods of natural pest control. Biological pest control involves the introduction of predator insects to keep pest in check. Ladybugs, for instance, can be used to control aphids, while praying mantids feed on a variety of unwanted insects. Other biological controls include:

•The strange-looking larvae of the green lacewing, which feeds on aphids, mealybugs, spider mites, thrips, immature scales, and whiteflies, as well as a many mite and insect eggs;

•The tiny Trichogramma wasp parasitizes the eggs of more than 100 species of insect pests, including cabbage worms, tomato hornworms, corn earworms, codling moths, cutworms, armyworms, cabbage loopers, webworms, and corn borers;

•Certain commercially produced bacteria are fatal to a number of pests. The bacillus thruingiensis, the bacteria that is being spliced onto the tomato seeds in California, is lethal to a number of leaf-eating caterpillars. One strain is effective against mosquitoes and black fly larvae in small ponds. The Milky Disease bacterium is reputed for killing a variety of grubs and is believed to have been responsible for the decline in the Japanese beetle population since the 1940s. As we mentioned above, until more is known about the long-term repercussions of introducing large populations of these bacteria into our environment, either through agricultural spraying or gene-splicing, caution should be exercised in using these products.

•Insect-killing nematodes (different from the plant-parasitic nematodes that damage crops, particularly grapes) are tiny organisms that are harmful to insect larvae living under or traveling on the earth's surface. They can be very effective against certain caterpillars and beetle larvae.

Chemical pest control involves the use of naturally occurring chemicals to control pests. Some of these chemicals are exuded by other plants, such as marigold, onions, and garlic. These plants can add color and variety to your garden, while naturally controlling pests. Other examples include:

•Certain scents can work to repel insects. These olfactory repellants include naturally derived plant oils from citronella, eucalyptus, cedar, and bay, all of which are effective against flying and crawling insects;

•Certain plants can also emit scents that will lure insects away from crops. These lures are often made up of components of the insects' favorite food and have been used effectively in traps for apple maggots, Japanese beetles, houseflies, and other species;

•Pheromones are hormonelike substances that act as sexual lures of pests. Pheromones are chemically isolated and designed to be used in traps for specific pests. They can be used against cherry fruit flies, codling moths, corn earworms, and flour beetles;

•Insecticidal soaps derived from nontoxic plant or animal sources can kill many pests when applied in water-based solutions. Unlike their synthetic counterparts, they are relatively harmless to beneficial insects like honey bees.

•There are now a number of botanical insecticides on the market that can be just as effective as their synthetic counterparts, but are safer and do not persist in the environment as long. These plant derivatives include rotenone, pyrethrins, ryania, and sabadilla.

We have barely scratched the surface of the alternatives available to conventional agricultural practices. There are a number of ways that you can get more information. First, of course, is your local bookstore or library for books on alternative agriculture or gardening. Remember, there may be a number of names under which this subject goes—"sustainable," "organic," "regenerative," or simply "alternative." Many health food stores also have gardening sections that may contain books on the subject. If you live near a farming community, you may want to check with local universities to find out what materials they have available and whether they offer any courses on sustainable agriculture. The University of California at Davis, the University of Illinois, and the University of Minnesota are all conducting research and offering classes. Also in Minnesota is the International Alliance for Sustainable Agriculture, a nonprofit organization headquartered at the University of Minnesota, 1701 University Avenue SE, Minneapolis, Minnesota 55414. This group, headed by founder Terry Gips, works both nationally and internationally to develop "economically viable, ecologically sound, socially just, and humane agricultural systems." The International Alliance's three major programs include research and documentation; organizational support and network building; and education and public outreach. They also publish a monthly newsletter called *Manna*. In the next section of this chapter we will discuss pesticides in greater detail and will refer to Terry Gips's book *Breaking the Pesticide Habit*, which is published by the International Alliance.

Conclusion

These are just a few of the facts concerning food production in America today. Despite the claims of agriculture officials and proponents of agribusiness boasting that the United States is feeding the world and our food has never been more efficiently produced, the real story of agriculture in this country is that it is becoming ever more vulnerable. Whether the world moves into the twenty-first century with a truly abundant and efficiently produced food supply depends in large part upon our willingness to see the current problems for what they are and to implement solutions to them. These solutions must be long-term and sustainable. Simply switching from one pesticide to another is not going to get to the root of the problem. "We must rethink not only our approach to pest control, but to agriculture and life as we know it," says Gips. Concerning the use of pesticides, Stuart B. Hill, professor of entomology at McGill University, says, "There are alternatives available. . . but first it requires changes in agroecosystem design and management. Alternatives are not alternative products, but alternative value systems and associated ways of thinking and behaving."

Merle Hanson, president of the North American Farm Alliance, suggests that individuals support bills that increase farm prices. The United States is the world's largest producer, exporter, and importer of food. We set the world price, and by keeping it too low, we are driving many Third World countries out of business, causing them to lose land and inhibit production–let your legislators know you're aware and interested.

For more information contact:

North American Farm Alliance
P.O. Box 2502
Ames, Iowa 50010
(515) 232-1008

Chapter 6
Pesticides

The farmer who phoned the USDA organic farming coordinator Garth Youngberg was distraught. He complained that his herbicides were just not working anymore. He spent about $45 per acre that year on chemicals that were ineffective on his 1,000 acre farm. Could he get some information from Youngberg about organic methods? And by the way, the farmer added, for the first time ever there was a sign posted at the local lake that said people were not permitted to swim there anymore because of the runoff from farm chemicals. He wondered whether his herbicides were strong enough.

The farmer's chemicals were very potent, certainly powerful enough to contaminate his community lake, but they were outmaneuvered by Nature. Herbicides kill weeds by preventing cell functions. But by the same process that enables any organism to survive by adapting itself against adversity, weeds alter their own constitutions to become immune to man made poisons.

Insects are also talented genetic engineers. Despite a tenfold increase in the use of pesticides between 1947 and 1974, crop losses due to insects have doubled, precisely because of chemical use. If a pesticide is strong enough, it will kill harmful insects along with their natural predators—birds and other insects. Pest insects are more numerous than predators because of population ratios, and are able to breed pesticide-resistant strains more quickly. Thus, killing off natural predators has helped increase the population of the pests.

A farmer's solution has always been to buy deadlier concentrates and use them at a rate that increases 15 percent each year. Between 1950 and 1970 the amount of pesticide needed to produce corn multiplied by 10 times, and necessary herbicides increased 20 times. In America, we use about five pounds of pesticide per person each year. Twenty percent of the billion pounds of pesticides applied annually in this country are used in populated, non-food-producing areas: factories, commercial settings, golf courses, and in homes.

To understand why such a dangerous and basically ineffective technology is still allowed requires a look at the methods of what has come to be called agribusiness. Farms have become food factories. The sole criterion is food production, not people protection. No consideration is given to the ecological relationships among humans, other animals, and plants. Attention is so fixed on production and profit that disruptions of natural balances are not acknowledged.

When a variety of crops is no longer planted and large tracts of land are devoted to growing one product, the need for diverse and specialized farming techniques disappears. Labor is cut to a minimum. Machines are designed to accomplish specific purposes. Such machines are economically feasible only if they can be applied to extensive acres planted with a uniform crop.

In such a context, the particular species of insect that feeds on the single crop is presented with an almost unlimited food supply, enabling the species to multiply greatly. At that point, the only defense available to agribusiness engineers is the chemical pesticide. Awareness that these chemicals will be useless in a year or so against the predictable arrival of genetically resistant insects does not act as a deterrent to spraying. For agribusiness, immediate profit is primary. In a year, agribusiness capital may be applied in a totally different field—mining or oil or transportation. Whatever ecological disruption is caused will be someone else's problem.

Or if the capital is to remain in food production, then the expectation of new chemicals to deal with the mutated insects gives the factory farmer the assurance needed to continue a single-crop approach. This addiction to synthetic pesticides has created a nearly billion-dollar industry that manufactures products that over the long run are ineffective and endanger our entire environment. Since we are on such intimate terms with these products—even the EPA admits that they are lodged in every one of our bodies to some degree or another—we ought to know something about them.

The Pesticide Explosion

It was a time of optimism. World War II had just ended. The high degree of technology achieved under the pressure of war would now be applied to a world at peace. Penicillin promised health where there was sickness. Nuclear power promised prosperity where there was poverty. There seemed to be no limit to the miracles science could produce.

In this context came the arrival of a most powerful agricultural tool: DDT. Its sponsorship could not have been more prestigious. Paul Mueller, a Swiss chemist working for J. R. Geigy (now Ciba Geigy) of Basel, was the first to recognize the insecticidal properties of DDT and was awarded the Nobel Prize for Medicine in 1948. Four years earlier *Business Week* magazine lauded the chemical ecstatically: "...mankind has new weapons promising eventual freedom from disease-bearing insects such as lice, fleas, flies, mosquitoes, and ticks; from household pests such as moths, cockroaches, and bedbugs, and from insects that frequently kill crops, orchards, and shade trees."

As late as April 1951, the *American Journal of Public Health*, citing country after country where DDT had impressively lowered the rates of typhus and malaria, editorialized about the presumed benefits and blessings of DDT: "This is one of the most dramatic and significant chapters in the the entire history of public health."

The irony of this statement did not become apparent until the publication of Rachel Carson's book *Silent Spring* eleven years later. The possibility that the entire planetary environment was being contaminated by indiscriminate use of these new pesticides had not been seriously considered in scientific research. Carson's book focused attention on an alarming set of data that indicated that whole species were in danger of being destroyed. Perhaps the human species itself would be wiped out.

The controversy generated by Carson led to investigation, and our most insidious pollution problem was uncovered. From an increasing body of facts, an ominous and frightening picture emerged. Because of its chemical stability DDT could travel intact through the food chain, concentrating itself in greater and greater quantities as it moved from level to level. DDT could be transported in water or water vapor; attached to particles of dust, it could circulate in the atmosphere. Not soluble in water, it did not dilute. Soluble in fats, it gravitated to biological units and through biological systems. Since it was deposited in the fatty tissues of man and other animals, time did not seriously impair its effectiveness as a poison. Even after 10 or 12 years, it retained 50 percent of its toxicity. Concentrated quantities of DDT lodged in the fatty tissue of fish, for example, could be lethal to birds that ate the fish.

Invisible, persistent, mobile, and dangerous, DDT spread throughout the entire biological community.

Rachel Carson cited a letter from a Hinsdale, Illinois, resident who described some of the results of spraying diseased elm trees in her town. After several years of DDT spraying to stem Dutch elm disease, the town was devoid of robins, starlings, and other local birds.

The destruction of birds in Hinsdale, Illinois, unfortunately, is not an isolated event. Dutch elm disease is widespread in the United States. It is known that this fungus is carried from tree to tree by elmbark beetles. Efforts to control the disease have mainly been limited to controlling the carrier insects. In town after town, especially in New England and in the Midwest where elms are profuse and highly valued, intensive spraying has become almost routine.

One such community is East Lansing, Michigan, home of Michigan State University. There, in 1954, two ornithologists, professor George Wallace and graduate student John Mehner, determined the intricate connection between spraying and the death of birds. The horror of dead and dying robins on the campus each day was a strong motivation for their work.

Although the users and manufacturers of insecticide were assuring everyone that sprays were harmless to birds, Wallace and Mehner could not help but suppose that the birds were, in fact, dying from insecticidal poisoning. The symptoms of such poisoning—loss of balance, tremors, convulsions, then death—could be seen in the dying robins. There were indications that the robins were not being directly affected by the chemicals, but indirectly by eating earthworms. A crayfish used in a research project accidentally ate an earthworm and died immediately. A laboratory snake was seized by violent tremors after being fed the worms. A key piece in the puzzle Wallace and Mehner were putting together was supplied by another researcher.

Roy Barker of the Illinois Natural History Survey, working in Urbana, unraveled the mystery of how the DDT connected with the robins to cause their death. Spraying the trees, which occurred in spring, created a poisonous and tenacious film over the surface of the leaves. This covering remained until autumn when the leaves fell, carrying their lethal cargo to the earth. The earthworms, performing their natural, ecological function, fed on the litter, gradually converting it into soil. Some of the worms may have died from the poison they ingested. The survivors acted as carriers. Barker determined that as few as eleven large earthworms can pass on a fatal dosage of DDT to a robin. A robin can consume about 10 to 12 worms in 10 or 15 minutes!

Citing such case histories and gleaning reports and details from various sources, Rachel Carson brought together in *Silent Spring* a mass of terrifying evidence. It wasn't just birds that were dying, but fish and animals as well. The poison was everywhere—in the soil and in the waters. It was in our food and it was in our drink. Nor were the effects of the poison limited to directly causing death. It was wiping out entire species by causing sterilization in the animals it did not kill. The bald eagle is a well-publicized example of a species decimated by pesticides. There is no lack of other examples of extinction accomplished and extinction threatened. Grebes, quail, pheasant, ospreys, falcons, even pigeons, are succumbing. And foxes, rabbits, raccoons, hawks, and blackbirds.

In addition to death and sterilization, Carson reported on studies made by Dr. Malcolm Hargraves of the hematology department of the Mayo Clinic. Dr. Hargraves, citing extensive clinical experience, "particularly during the past ten years" (before 1962), contended that the "vast majority" of patients suffering from blood and lymph disorders had been extensively exposed to DDT and other newly developed pesticides generally of the hydrocarbon group. Reports from Sweden, Japan, and Czechoslovakia provided more evidence of linkage between the pesticides and leukemia. Contamination was shown to be rampant and continually increasing. And the results of it were overwhelming.

All of this would have been quite enough, but Carson also showed that contamination was not the only problem that discredited the new pesticides. Another one—genetic resistance—made the whole matter even more appalling. In 1943, the Allied military government in Italy dusted great numbers of people with DDT and thereby waged a successful campaign against typhus. Encouraged by this achievement, extensive spraying was utilized two years later with the aim of controlling malaria mosquitoes. Only a year following this spraying, one genus of mosquito had developed resistance to the spray.

Among most experts and investigators, it has been proved that chemical-pesticide technology was at best an unworkable sham and at worst a dangerous environmental threat. Yet the production of these chemicals and the use of them continued on a large scale.

In 1969, the Department of Agriculture finally prohibited the use of DDT on shade trees and tobacco plants, in marshes and estuaries, and in households and gardens. Up until 1972, when it was banned for most uses by the EPA, 1.2 billion pounds of DDT were used by the departments of Agriculture and Interior to exterminate forest insects, by the military to combat bats and mice, and by farmers growing cotton, beans, sweet potatoes, peanuts, corn, and other crops. Despite this ban, which followed a bitterly fought battle between environmentalists and proponents of the chemical's use, the Department of Agriculture continues to foster the use of other persistent insecticides. There has been off-again, on-again banning of DDT by the Food and Drug Administration and, as recently as 1985, the EPA was still equivocating on whether or not to ban pesticides such as Difocol, which typically contains DDT. Discoveries of DDT were just the beginning. Here is a closer look at some of the chemicals that are poisoning the food chain.

The Dirty Dozen: Case Studies in Poison

The most deadly in man's arsenal of pesticides are commonly lumped into twelve groups, hence their sobriquet "the Dirty Dozen." The groups, together with some of their trade names, are as follows:

1. *DDT:* This well-known pesticide goes under as many as 65 different trade names. Some of these are Antelope, Cotton Dust 3-10-0 (also 30-10-40 and 3-9-0), Difanil, Dicophane, Dinoside, Pentech, Zerdane, Double Swallow, and Cesarex. Difocol is a pesticide that is manufactured from and virtually identical to DDT. Manufacturers technically insist that they do not need to list DDT on the label because they claim it is a "contaminant," not an ingredient. Consequently the one to 20 percent concentrations of DDT found in difocol falls outside the regulatory structure.

2. *Chlordane and Heptachlor:* Trade names for Chlordane include Belt, Endrinet, Gold Crest C, Gold Crest Termide, Octa-Klor, Termide, Syndane, Velsicol 168, and Velsicol 1068. Heptachlor goes under some of the following names: Agroceres, Drinox, Cold Crest H, Hold Crest Termide, Terra San, Termide, and Velsicol 104.

3. The *Drins.:* aldrin, dieldrin, and endrin. Aldrin's names include Aldocit, Aldrex, and various other "Ald" derivatives, Drinox, Octalene, and Seedrin. Dieldrin has its own name derivatives such as Dieldrex, Dieldrite, and Dielmoth, together with Red Shield, Permetezo, Termitox, and Pestex. Endrin is marketed under "Endr" names like Endres and is also sold as Mendrin, Insectrin, and Hexadrin.

4. *EDB:* or ethylene dibromide was first introduced under the name of Bromofume. While still going under that name, it is also called Celmide, Soilfume, Nephis, Pestmaster, and Fumo-Gas.

5. *Chlordimeform or CDF:* most commonly known under the names of Galecron and Fundal. Other names include Ciba 8514, Ciba-C 8514, Ovitix, Spanon, Schering-36268, ENT 27335, and ENT 27567.

6. *DBCP (Dibromochloropropane)* : is also known as Nemafume, Nemagon, and other "Nema" derivatives, Fumagon, and OS 1897.

7. *2,4,5-T:* Some of the commercial names of 2,4,5-T include Brush Killer, Fruitone, Weedone, Weed Be Gone 21, and Envert-T. It also goes under a number of French and Spanish names targeting developing countries: Herbicide Mataarbustos, Debroussaillant Concentre, and Mata Arbustos Baja Volatilidad 21.

8. *Lindane (BHC) and Hexachlorocyclohexane (HCH):* These two chemicals are closely related, the difference being that while HCH is a complex of isomers, lindane is primarily made up of the most deadly of these called gamma-HCH. Commercial names of these two cousins include derivatives of "lin" and "gamma" such as Gamma-Col, Gammahexa, and Gamacid, Lindaterra, and Linda-tox, together with others such as Agrocide2, 6G, 7G, and Nexit.

9. *Paraquat:* best known for its use in destroying marijuana crops in Latin America, is marketed under some of the following names: Weedol, Actar, Gramoxine, Pared, Pathclear, Sweep, and Simpar.

10. *Parathion:* This phosphate-based pesticide is commercially marketed under some of the following names: Folidol, Ekatox, Sixty-Three Special E C, Vapophos, Niran, and Pestox Plus.

11. *Pentachlorophenol or PCP:* This potent insecticide and fungi-cide is one of the most commonly found contaminants in the na-tion's Superfund sites. Its commercial names include Penta Dragon 50 Pino, and other "Penta" names like penta nol and Penta Ready, and Term-i-trol and Chem-Tol.

12. *Camphechlor:* most commonly known as Toxaphene. Its other names include Duo-Tox, Toxadust, Motox, and Chem-Phene.

DDT

Most modern pesticides fall into two categories: the organic chlorines and the organic phosphates. DDT was the first and the most widely applied of the organochlorine compounds. This now-infamous chemical is the quintessential pesticide, and its story closely parallels those of the other chemicals used to combat na-ture's "pests."

DDT is a hydrocarbon derivative which, means that its production is linked to two of the world's largest and most influential multinational conglomerates—the pharmaceutical and oil industries. With little regard for its side effects, DDT was almost immediately and unanimously proclaimed a miracle chemical. When the myth crumbled, use persisted and in fact increased. Rather than solving our pest problems, it has compounded them with the development of stronger mutant strains. In the meantime, DDT has greatly decreased populations of natural and beneficial predators and has added greatly to environmental contamination.

The myth of harmlessness that surrounded DDT for a long time was probably due to the dusting of soldiers, prisoners, and refugees during World War II to combat body lice. Although millions of people were dusted with the chemical, there were no reports of immediate negative effects. With small amounts of DDT, dangerous effects are delayed. In concentrated amounts, as in aerial crop dusting, pilots receiving a sudden blast by an abrupt windshift have died instantly.

The body-lice application of DDT caused no immediate harm because the chemical is not readily absorbed into the body in powder form. Only when dissolved in oil, as it is for most uses, including agricultural spraying, does it become toxic. DDT can accumulate in the body from exposure through a number of routes. Eating contaminated foods permits intestinal absorption; breathing airborne DDT-laden aerosols, vapors, and dust introduces the chemical through the lungs; and touching contaminated objects allows direct absorption through the skin. DDT lodges in fatty tissue and accumulates there. It is especially partial to the liver and kidneys. Minute quantities can bring about grave malfunctioning of the human body. For example, three parts per million (ppm) has been shown to inhibit the action of the heart muscle, and five ppm causes disintegration of liver cells. It is cumulatively stored in the body and because of its persistent quality threatens eventual chronic poisoning.

In a study performed at the Environmental Health Center in Dallas, 200 patients known to be highly sensitive to synthetic chemicals were tested for common chlorinated–hydrocarbon pesticides. DDT was found in the blood of 62 percent of the patients. Each part of DDT observed in a person's blood is known to correspond to 306 parts lodged in the fat tissues, 27 parts in the liver, and 6.5 parts in the brain.

DDT has become so widely disseminated in food that such poisoning can take place as a result of a simple and normal diet. Due to its long life and toxicity, scientists have estimated that we have experienced to date only 1 percent of the damage that will occur as a result of the DDT already used. In other words, even if the use of this chemical were to be totally discontinued today, 90 percent of the damage it does would still lie ahead of us.

Chlordane and Heptachlor

These two related chemicals are widely used organochlorine insecticides. A former administrator of the FDA described chlordane as being so toxic that "anyone handling it could be poisoned." Unfortunately, this warning has often gone unheeded. Chlordane has been liberally sprayed over residential lawns and golf courses throughout the country and has widely contaminated our food supply. Heptachlor has been detected in most species of Arctic wildlife and has been linked to widespread milk contamination in Hawaii. The contamination resulted from the widespread use of heptachlor in spraying Hawaiian pineapples. Both chlordane and heptachlor have been applied to a wide variety of crops, including corn, cotton, pineapples, strawberries, apples, tobacco, and citrus. For home use they have been promoted to fumigate soil, to control crabgrass, fleas, mosquitos, and termites. Like DDT, these two chemicals are not easily biodegradable and can persist in the environment for up to 20 years after application.

Chlordane and heptachlor also share DDT's characteristic of accumulating in fatty tissues. They are also capable of crossing the placenta, and studies have shown that fetal blood can accumulate from two to four times as much of the toxic chemicals than the mother. Both chemicals are suspected of causing a wide range of debilitating disorders ranging from cancer to brain tumors in children to liver dysfunction. Chlordane accumulates particularly in the digestive tract and thus is linked to stomach cancer. Chlordane is very toxic when absorbed orally, but can be even more dangerous when taken in through the skin or inhaled. Heptachlor is considered 6 times more toxic and 30 times more volatile than chlordane. Common symptoms of exposure to these two chemicals include dizziness, respiratory problems, and nervous–system disorders that can alter brain-wave patterns, and cause insomnia, nausea, and tremors.

Chlordane was introduced in 1945 and heptachlor in 1948. As early as 1958, countries around the world started to issue limitations or bans on their use. It was not until 1975 that the United States issued restrictions on the use of chlordane, while we waited until 1978 to ban heptachlor for all uses other than termite control. Even these restrictions proved ineffective at eliminating the risks associated with the two potent organochlorines. In April 1987, an EPA study found that chlordane, a pesticide used in millions of homes to control termites, posed a high risk of cancer even when properly used. Estimates suggested that when used in a certain type of construction, the chemical could contaminate the air and result in risks between 1 and 3 in 1,000 of developing cancer over 70 years of exposure. Although one major manufacturer's spokesmen contended that their "risk data does not show a problem" and that the EPA statistics were "erroneous," in August of that year, the company agreed to stop selling both chlordane and heptachlor immediately. In the meantime, during the preceding 40 years, more than 24 million homes across the country had been treated with similar chemicals for termite extermination. Moreover, the agreement reached by the manufacturer and the EPA still did not prohibit the use of existing stock that distributors and pest-control companies already possessed. This maddened environmental groups. "It's a travesty that the EPA will allow people to be exposed to this poison while existing stocks are being used," said Cynthia Wilson, executive director of Friends of the Earth. That group and several other environmental associations, filed suit in federal court seeking an order to force the EPA to issue an emergency "stop-use" order on the chemicals based on the evidence of health risks.

The concern of the environmental groups proved wellfounded, for in March 1989, almost two years after the EPA report, serious health risks were still being reported. One of the biggest poultry producers in the country had to destroy 400,000 chickens because they were contaminated with a cancer-causing pesticide that had been banned 11 years earlier. The pesticide was heptachlor. The damage to Arkansas' $3-billion poultry industry was considered the most serious incidence of pesticide contamination in that state's history, and was blamed by food producers on the federal government's ineffective regulation of dangerous agricultural chemicals. As in the case with the later ban on the use of the chemicals for termite fumigation, when the EPA banned the use of heptachlor, the agency allowed the continued use of existing supplies of the chemical—enough to last for 80 years. Although the EPA has banned heptachlor for use in food production, it continued to be allowed to treat seed grains. Coupled with regulatory

inefficiency, this ambiguous distinction is most probably what led
to the poultry contamination. While nobody knows for certain
how the chickens came to accumulate such high doses of the pesticide, investigators at the USDA and the FDA believed the contamination was caused by sorghum seeds sprayed with heptachlor and
sold illegally as feed grain. As another example of regulatory failure, the contamination was not discovered by the agencies responsible for policing the safety of our food supply, but by the Campbell
Soup Company, which had purchased the meat for processing.

Nor is this poultry contamination an isolated event. In early
1986, heptachlor was discovered in livestock feed that was being
widely distributed to dairy farms in eight midwestern states.
Dangerous levels of the carcinogen appeared in the cows' milk,
resulting in the shutdown of nearly 100 dairy farms in Arkansas,
Missouri, and Oklahoma, together with the recall of tens of thousands of gallons of tainted milk.

Wildlife groups have also discovered that waterfowl migrating
along the flyway transversing the Great Plains were contaminated
with 18 pesticides, including heptachlor.

The Drins: Dieldrin, Aldrin, and Endrin

This deadly trio, which also belongs to the organochlo-
ride family, first made its appearance in the early 1950s.
From the outset, some 80 percent of all aldrin in this country was
used to treat cotton. The remainder was used for a variety of crops,
including bananas, potatoes, sugar beets, fruit trees, tobacco, and
sugar cane. Both dieldrin and endrin were used on a range of other
crops, such as cocoa, cereals, coffee, tea, and a number of vegetables. Again, many countries around the world started banning
or restricting the use and distribution of the three drins during the
1960s, but it was not until 1976 that the United States took any
action to control their use.

All three of the drins have been shown to cause all sorts of adverse health effects in animals: cancer, particularly of the liver,
spontaneous abortion, birth defects, and brain and nerve damage.
Aldrin readily mutates in nature, and is metabolized by most animals into dieldrin. So while aldrin does often not appear in its own
chemical configuration, according to a 1971 survey sample, 99.5
percent of the population had detectable traces of dieldrin stored
in their fatty tissues averaging about .29 ppm. Due to accidents
with pesticide overexposure, we now know that dieldrin is a potent poison to both the brain and nervous system, capable of almost
immediately causing seizures and convulsions. It is estimated to

be 40 times more toxic than DDT when absorbed through the skin. The World Health Organization reported that when Dieldrin was substituted for DDT in areas where mosquitoes had developed resistance to DDT, cases of severe poisoning occurred among the sprayers themselves. All of them experienced convulsions. Several died. And some survivors experienced convulsions as long as four months after exposure.

In Western Australia, where dieldrin is legally used to treat homes for protection against white ants, the breast milk of 140 nursing mothers was found to contain an average of 14 times the maximum acceptable intake of the substance as set by the World Health Organization.

Aldrin is not only lethal but also causes sterilization. A small quantity about the size of an aspirin tablet is enough to destroy 400 quail. In less lethal doses, it has caused pheasants to lay eggs that will not hatch. Rats exposed to aldrin have a lowered fertility rate, while those who do conceive bear sickly offspring that do not live very long. Puppies born of mothers treated with this chemical died three days after birth. Pigeons fed dieldrin have developed enlarged thyroid glands. When ingested by mammals, Endrin is metabolized into three compounds, each of which is even more harmful than the endrin itself. This chemical is so toxic that it has been reported as causing illness and death even when great care was used in its handling.

When the Montana Department of Fish, Wildlife, and Parks investigated contamination of the state's waterfowl, which feed on pesticide-laden grasses, it discovered high levels of Endrin, DDT, and 17 other chlorinated hydrocarbons that were up to ten times the allowable levels for domestic poultry. The findings prompted the Montana's Department of Health to recommend that the waterfowl— hunting season be canceled in 18 Western states. One of the officials called the birds "flying garbage cans full of pesticides."

The Organophosphates

The second group of modern insecticides consists of the organic phosphates. These substances were first recognized for their insecticidal qualities by a German chemist, Gerhard Schrader, in the late 1930s.

These insecticides strike at the nervous system in much the same way as deadly nerve gases. Impulses in a normally functioning nervous system are passed from nerve to nerve through a chemical transmitter called acetylcholine. The existence of acetylcholine in the body is extremely transient. It vanishes as soon as it is used. Otherwise, the impulses it carried would continue to flash from nerve to nerve, resulting in lack of body coordination, tremors, spasms, convulsions, and eventually death. The body is protected against this by an enzyme called cholinesterase, which acts to destroy the acetylcholine once it is no longer needed.

Organophosphates destroy cholinesterase on contact. If enough of the chemicals are present, enzyme levels in the body are reduced, and the entire nervous system is thrown into disorder.

Parathion, one of the first of the organophosphate pesticides, was first introduced in the late 1940s in answer to the resistance that insects were already developing to the organochlorines. It soon became one of the most widely applied pesticides in the world, being used on much of our grains, fruits, vegetables, cotton, tobacco, and nuts. This lethal substance is responsible for hundreds of deaths each year. In California alone, millions of pounds of the pesticide are spread over citrus groves by hand, by motorized blowers, and by aerial spraying. The chemical is so potent and persistent that in Riverside, California, orange pickers suffered parathion poisoning when they worked groves that had been sprayed almost three weeks earlier. According to Robert Metcalf, a scientist who has studied the effects of the pesticide, "Parathion is widely used to control insects but is a general biocide and has no place in pest management. . . and is too toxic for general use." The National Institute for Occupational Safety and Health estimates that 400,000 people are exposed to the toxic properties of parathion in the course of its manufacture, shipping, storage, and application. Its residues have been found to persist up to 16 years.

Among farmers, Temik (or aldicarb) is one of the most popular pesticides ever developed. Young plants absorb the substance through their roots when it is mixed with topsoil or sprinkled on the ground after they emerge. The plants then distribute it throughout their vascular systems. By eating any part of a Temik-dosed plant, an insect takes in enough of the pesticide to induce fatal nervous–system disorders. Various impairments of nerve functions are also experienced by people who are exposed to Temik. Muscle spasms, nausea, convulsions, and death have resulted from exposure to large amounts, while laboratory studies suggest a connection between repeated small-dose exposure and cancer, neurological disorders, and miscarriages. Temik's composition is based on MIC, the chemical that leaked from a Union Carbide plant and killed 2,000 people in Bhopal, India, in 1985.

2,4,5-T and Dioxin

Innocuous as its numbered name may sound, this potent herbicide, together with the similarly named 2,4-D, are what make up Agent Orange, the controversial defoliant used during the Vietnam War. Also related to these numbered toxins is 2,3,7,8-tetra-chlorodibenzo-p-diozon, commonly known as dioxin, a highly deadly substance that is both formed during the manufacture of the herbicide 2,4,5-T and afterward can be found as an "impurity." All of these chemicals are extremely toxic and very pervasive. Even people who have not knowingly been exposed to dioxin, for instance, have been shown to have significant contamination, says Dr. Arnold Schecter, professor of preventive medicine at the State University of New York, Binghampton.

Carol Van Strum, a mother of four, has experienced firsthand just how dangerous these chemicals can be. One day her children were down playing by a river near the house. They stumbled home with burning skin, mouth, throats, noses, and eyes; by nightfall, they were ill with nausea, vomiting, headaches, and diarrhea.

During the weeks that followed, Carol noticed that many plants died in her own garden as well as in the surrounding area. Chicks, geese, and ducks hatched on her property with deformities. The family dog's back legs became paralyzed. When the family revisited the river whose banks were sprayed, they were appalled to find an oily scum on the water, dead fish, snails, and birds as well as trees with twisted, unnatural growth.

Assuming that a dreadful mistake had been made, Carol and her husband called the county road department. They were assured that what had been sprayed was only the "perfectly harmless" 2,4,5-T and that they could "drink the stuff without being hurt."

Within a month, U.S. Forest Service helicopters sprayed the same substance along a nearby ridge. Children of other families became ill, mothers miscarried, livestock aborted and lost their hair, poultry died, and crops withered. Carol and her family were sick again too, with a new complication—uterine bleeding by both Carol and her daughter, who had not yet begun to menstruate.

The Van Strums attempted to alert the local health department, fish and game officials, and the EPA, which referred their letters to the U.S. Department of Agriculture. Their answer claimed that no studies had ever shown 2,4,5-T to have the effects that the Van Strums described. The Dow Chemical Company sent a lavish brochure on phenoxy-based herbicides, the class to which 2,4,5-T belongs. The brochure was illustrated with color photographs taken by a Dow associate scientist, Eugene Kenage, who was also president of the Michigan Audubon Society. In a personal letter Dr. Kenage assured the Van Strums that the herbicide was safe.

Carol's husband, Steve, went to the state university library, where he discovered studies on 2,4,5-T that did not agree with Dow or the USDA. In the first place, the compound made up 50 percent of Agent Orange. Steve found many reports that documented events following herbicide treatments which were similar to what happened on his farm. He discovered that the herbicide affects any cells—plant or animal—that it destroys immune systems, and wreaks havoc with the genetic material—all effects that can cause cancer and mutations.

Steve and Carol wrote letters to congressmen, senators, and state officials, but they were either referred to the USDA or answered with the USDA statements they had already heard but no longer believed. Two years later, however, the Van Strums and other local residents successfully sued the U.S. Forest Service and were able to prevent the use of dioxin-contaminated herbicides such as 2,4,5-T in their area.

Ethylene Dibromide or EDB

This potent pesticide of the fluorinated–hydrocarbon family was first introduced by Dow Chemical in 1948 under the name of Bromofume. Its primary agricultural use is as a postharvest fumigant for grains, fruits, and vegetables. Although its dangers have been clearly established, EDB is still only under minimal regulation by the federal government. The National Audubon Society claims that it "poses the highest risk of getting cancer" of any known chemical. Even the often-conservative EPA agrees. It has been shown to cause tumors in laboratory animals, even following extremely short exposures, and evidence suggests that the chemical also causes birth defects, miscarriages, and other health problems.

Nearly 300 million pounds of EDB are manufactured each year in the United States. Ninety percent of it is used as an antiknock additive in gasoline to prevent lead deposits from collecting in engine cylinders. Until 1984, most of the remaining 10 percent were injected into soil to protect crops against root worms. The rest of the EDB, at least 1 million pounds per year, was used to fumigate stored grains, fruits, and vegetables. According to an EPA sample of American grains in the early 1980s, nearly 60 percent was found to contain detectable levels of EDB. When the FDA conducted surveys of harvested fruits, it reported to the EPA in October 1981 that virtually all fruits sampled had EDB residues.

The primary objective of the FDA in its surveillance programs is locating and removing from commerce food and feed commodities that contain illegal residues. The problem is that what constitutes an *illegal* residue is often determined by either the FDA or the EPA or both. In order to define an illegal residue, tolerance levels are set by the agencies. Often what happens is a Catch-22 situation in which the agencies use their own limitations as an excuse for not doing something that they do not want to do.

What becomes all too apparent is that without good faith, government agencies can hide behind bureaucratic technicalities and use the law to subvert rather than protect the best interests of the public. It is alarming to realize that almost as soon as one chemical is banned, industry has moved to replace it with another similar and possibly even more dangerous one. Without a government commitment to protecting the public, this cat-and-mouse game can go on *ad infinitum*. In the end, the American public is the loser as it continues to finance all of this government rigamarole and still does not end up with a safer food supply.

Since Rachel Carson's *Silent Spring* was first published in 1962, pesticide use has more than doubled; while she wrote of some 200 of these chemicals, today more than 600 are in wide use. By 1989, the pesticide scare spread across the nation. "Farmers and food companies across the country," reports *The New York Times*, "are taking expensive new measures to test for pesticide residues and convince consumers that fruits and vegetables are safe to eat." Maine potato farmers, vineyard owners in California, and a New England supermarket chain are only a few of those who have taken steps to make the food industry chemically safer.

Every year modern pesticide technology creates hundreds of new compounds. Many of them spring up to replace toxic substances that have been taken off the market only after the arduous, expensive, and time-consuming efforts of conservationists and consumers. Chemical companies know that it can take years to ban a toxic substance and are happy to reap financial benefits from a product until forced to withdraw it from the market.

Until recently even after a chemical was banned or its uses restricted, the chemical industry's influence did not cease. In many cases, bans would only take effect a number of years after the completion of the review process. The reason for this was to allow chemical manufacturers to unload all existing supplies of the chemical, so that they would not suffer any economic loss due to the ban. In the event that the chemical was banned outright, the law worked even more favorably for the manufacturers, requiring the EPA to buy up all existing stocks of the pesticide at *market value* and to assume the costs of disposing of the hazardous substance. As an example, when the EPA banned the pesticide Dinoseb, the agency was forced under the law to buy up and dispose of all existing stocks of the chemical at a cost of more than $100 million. The agency's total budget for pesticide control is around $60 million. Toxicologist Janet Hathaway estimates that if the EPA immediately removed just seven of the most dangerous and widely used pesticides, indemnification costs could be more than $400 million. The EPA's estimate of the cost of removing the widely used soybean and corn herbicide alachlor, identified by the National Academy of Sciences as posing a significant cancer risk, is $161.5 million. Commenting on the effects of the law, Hathaway says: "No other law promises federal funds to companies that manufacture extraordinarily dangerous products. Even politically powerful industries like automobile and pharmaceutical manufacturers must absorb the costs of their mistakes and recall defective products with no governmental assistance at costs incurred themselves. An article in the *Los Angeles Times* summed up the problem: "Pesticide indemnification and disposal obligations pervert incentives for companies to develop safer pesticides or to withdraw agricultural chemicals as soon as hazards are identified...these requirements effectively eviscerate EPA's power to halt the use and sale of pesticides that are found to pose grave risks."

In September 1988, it began to look as though things were going to change. The House of Representatives approved a bill to speed up the review process of these hazardous chemicals and set a 9-year deadline for the retesting of some 600 active ingredients used in 50,000 commercial pesticides. One week later the same bill was adopted by unanimous consent in the Senate. Similar attempts over the years to amend the pesticide act has been repeatedly thwarted by powerful lobbyists for the chemical manufacturers, farmers, and food processors.

The bill also proposed long overdue changes affecting the financial side of pesticide regulation. The first change was a revocation of the infamous indemnification clause that not only was an unconscionable burden upon the taxpayers, but also restricted the EPA's ability to remove hazardous chemicals from the marketplace.

The second change proposed in the new legislation was a shift in the cost of retesting pesticides. Previously even when the original testing of a chemical proved to be defective or flawed, it was the agency and not the manufacturer who bore the costs of retesting. Under the bill, these costs, ranging from between $50,000 to $150,000 per chemical, would be borne by the manufacturer. This change is particularly significant because of the large degree of research fraud that goes on in the testing of pesticides. In 1983, the EPA reported that two-thirds of all tests performed to establish the safety of 212 pesticides on the market were scientifically invalid. Four officials from the large research company hired to test the safety of the chemicals were indicted by a grand jury on federal criminal charges of misrepresenting test data and supplying false lab reports. The EPA, however, did nothing about the chemicals, many of which are still in wide use. Although an EPA spokesman conceded that the records of the research company, Industrial Bio-Test, were "shocking," agency officials maintain that they do not have the authority to ban invalidly tested chemicals.

Under the gun from Congress, the EPA decided in October 1988 that it would make some changes in its regulation of pesticides. Even these changes strongly bore industry's mark and clearly showed how little the agency could be trusted to act in the public's best interests. The changes announced by the EPA were designed to reduce the overall threat of cancer to consumers. They would be affected by closing some of the loopholes of the 1972 law—but only at a price. While the agency plans to apply a uniform standard to all pesticides—regulating residues of both new and old cancer-causing pesticides in both processed and raw foods—the standard would be weaker. The result, according to John A. Moore, the agency's acting deputy administrator, will be to remove some of the more dangerous older food pesticides from the market, but allow newer pesticides that are thought to present a lower risk. Critics disagree. The 1972 law disallowed contamination by any cancer-causing pesticides. Instead of closing the loopholes in the law and expanding its provisions to cover both pre-1972 chemicals and raw food, the agency saw fit to relax the standard to one of a "negligible risk" for all pesticides in all foods. This negligible–risk standard is defined by the agency as a one-in-a-million chance of getting cancer over a 70-year lifetime.

The problems with our current system of pesticide regulation do not stop there. While those setting tolerance standards for pesticides would like the public to believe that an objective science of risk assessment exists, the truth remains that assessing the health risks of a given chemical is a highly subjective and often politically charged issue. How much of a suspected carcinogen does it take to make drinking water unsafe? The exhaust of garbage incinerators contains minute levels of dioxin, but how much can a neighborhood reasonably tolerate in its air?

Another problem that has long bothered environmentalists in the agency's risk–assessment process is its failure to take into account differences in exposures and sensitivities to chemicals. The Natural Resources Defense Council (NRDC), a private organization specializing in health and environmental issues, is especially concerned about the EPA's failure to take into account the health risks for children. A two-year study conducted by NRDC has found that there is a greatly increased risk to children from chemicals in their diets. According to the study, young children are at an especially high risk because they eat more fruits and vegetables per body weight than adults and are more sensitive to chemicals. The study estimates that 5,500 to 6,200 children in the current preschool population "may eventually get cancer solely as a result of their exposure before 6 years of age." According to the report, this represents a risk of cancer as 5 in 20,000, much higher than the "one-in-a-million chance" allowed by the EPA.

The NRDC cites a number of factors undermining the validity of the EPA's tolerance levels. Tolerance levels are set for an average American adult of 60 kilograms (about 132 pounds). This does not take into account different tolerances for children, pregnant women, chemically sensitive individuals, the sick, or the elderly. The EPA claims to account for these differences in the calculation of what is called the "Safety Factor." The NRDC, however finds that tolerance levels set by the EPA "are based on tradition, not science."

Here is how the Safety Factor is calculated. First, starting from data provided by the chemical manufacturer, the EPA determines the NOEL (No Observable Effect Level). That is the level at which none of the adverse effects (cancer, nervous disorders, etc.) showing up at higher dosages are observed in laboratory animals. The Safety Factor is obtained by multiplying the NOEL by 100. The figure of 100 is based upon the EPA's assumption that humans are 10 times as sensitive to toxic substances as animals and that this susceptibility can vary another 10 times among humans.

Even the EPA's own Science Advisory Board has admitted that "the scientific data supporting this practice [are] scanty at best. . . . No data have been produced to support either the extrapolation factor of 10 to cover the sensitivity of humans compared to the test species or to cover the range of human sensitivity."

Another problem with the EPA tolerance calculations is that they fail to take into account the synergistic toxic effect of chemicals. This occurs when the interaction between two chemicals can produce even more toxic substances than either of the chemicals individually. Most crops treated with agricultural chemicals are subjected to a number of different chemicals. The soil may be fumigated, chemical fertilizers applied, the seed itself may have been treated. And the plant and its fruit may be sprayed a number of times with pesticides while it is growing. Upon being harvested, the crop may be treated with fumigants a number of times before being sold and processed into food. During the processing stage, dyes, emulsifiers, and preservatives are compounded with these pesticide residues. The EPA generally ignores the fact that pesticide residue levels on crops are sometimes made more dangerous as they pass through the various steps of food processing, cooking, and digestion.

In order to account for the dietary differences between individuals, the EPA applies what is called the "Food Factor." This is a rough estimate of the amount of a food that a person eats each year based on the Department of Agriculture's statistics. The numbers used by the EPA are either the 1975 statistics on total food production (minus an estimated amount for loss and spoilage), or figures from a 1965/66 household–consumption survey. These numbers are then divided by the U.S. population to give the Food Factor for each commodity.

Even if the production or consumption statistics used as a base were current, the resultant calculations would remain extremely speculative. When the shifts in dietary patterns during the past 10 to 20 years is taken into account, these crude guesses become meaningless. People, for instance, are eating much less red meat and more chicken and fish. Fruit and vegetable consumption has increased significantly, as has the use of whole grains. "The EPA," says the NRDC report, "assumes we eat no more than 7.5 ounces per year of artichokes, avocados, blueberries, eggplants honeydew melons, nectarines, or radishes. Therefore, if consumers ingest more than 7.5 ounces per year of any of these commodities, they are potentially exposed to more pesticide residues than EPA considers safe." Other foods that the EPA assumes to be eaten in quantities less than 7.5 ounces in one year include mushrooms, parsley, walnuts, almonds, figs, garlic, mangoes, Brussels sprouts, cantaloupe, and tangerines.

Cancer-causing Chemicals Remain in Wide Use

In 1980, the EPA ordered a special review of how much of the pesticide captan it should allow on fresh fruits and vegetables.

The EPA proposed a ban on captan in 1985. But in 1989, instead of a ban, the agency proposed revising the allowable level—called the tolerance—for captan residues on certain crops, and fell far below what agency critics believe is necessary.

And so the story goes.... Captan is no exception. Nor was DDT, EDB, Alar, or any of the other highly toxic pesticides that are recognized carcinogens and yet remain an integral part of our food supply because of EPA inaction. Particularly frustrating is the EPA's open admission of the dangers of pesticides. The agency itself ranks the chemicals as the nation's number three cancer risk following worker exposure to chemicals and indoor radon. But still the agency does not act. *Why?* John A. Moore, the acting deputy administrator, says that the agency simply lacks the funds and the resources to deal with the problem. The agency also refused to ban the chemical because "current data don't warrant a ban." These excuses for bureaucratic inaction recur time and time again with infuriating regularity. Below are some examples:

• In 1984, the EPA began a special review to determine whether the pesticide alachlor should be banned. Manufactured and sold by the Monsanto Company under the trade name of Lasso, alachlor is one of the nation's most widely used herbicides. About 80 to 84 million pounds of the chemical are used each year.

Alachlor was linked to cancer in laboratory animals, and one agency official recommended the immediate "emergency suspension" of the chemical when it began showing up in high concentrations in tests of drinking water across the country. In December 1987, the EPA issued a statement confirming the presence of alachlor in drinking and surface waters "has been identified as a cause for concern." The statement also said that people could be exposed to the pesticide by eating foods having been sprayed with the chemical and that farmers using the chemical faced a relatively high risk of cancer. In the end, however, the EPA concluded that the risk of getting cancer from alachlor was "generally" one in a million and that this "reasonable" exposure did not warrant banning the chemical.

• In its report on pesticides in 1987, the National Academy of Sciences (NAS) found that one third of all fruits and vegetables grown in this country are treated with a class of fungicide called ethylene-bisdithiocarbamates (EBDCs). Thirty million pounds of these chemicals are used each year on one half of all our potatoes, one third of all oranges, apples, tomatoes, and spinach. Other crops are also heavily treated. According to Richard Wiles, co-author of the NAS report, "For dietary exposure through pesticides they are about the worst."

EBDCs have long been recognized as causing cancer and birth defects. They have been under special review by the EPA since 1977. After studying them for five years, the EPA announced in 1982 that it did not have sufficient information to warrant banning the chemicals and returned them to the market for general use with a warning label. It was only under pressure from a lawsuit instituted by the AFL-CIO that the EPA agreed to reexamine EBDCs. The chemicals were put back under special review in July 1987, but the EPA extended the deadline for its decision to ban EBDCs until 1990—a full 13 years after it undertook its initial review of the chemicals.

The War Is On

   On August 14, 1988, Cesar Chavez, the controversial president of the United Farm Workers of America, was on the 29th day of a water fast. In 1969, Chavez had orchestrated one of the first successful food boycotts when he convinced 12 percent of the American population to stop eating table grapes until growers would meet farm workers and negotiate over working conditions. Again, almost 20 years later, another fast protested working conditions for the farm workers, but this time the issues were very different; and this time, the issues affected all of America. Concerned for risks of cancer and other illness caused by working with and around hazardous agricultural chemicals, Chavez's message was to not buy California grapes until growers agreed to ban the five most dangerous pesticides currently used in grape production. Especially troubling was a cluster of cancers and birth defects showing up in communities around the San Joaquin Valley. In the small town of McFarland, six children had recently died.

   Among the chemicals Chavez hopes to see banned are:
   •one fungicide under review by the EPA since 1985 as a suspected carcinogen;
   •an herbicide and pesticide that was officially banned by the EPA after it was found to cause both birth defects and cancer. The chemical is, however, still allowed until growers can find an alternative;
   •methyl bromide, the fumigant that sprang into wide use to fill the gap left by EDB. This chemical is responsible for more occupa-

tional deaths than any other pesticide used in California.

•parathion and phosdrin are two organophosphate pesticides that operate by attacking the nervous system of insects. Due to its extreme toxicity—one teaspoon spilled on the skin can be lethal— the EPA restricted the use of parathion to certified, trained applicators. Because of parathion's toxicity when inhaled or touched, many believe that the EPA precautions provide little protection for farm workers.

Obviously, the 61-year-old Chavez takes the issue of pesticide regulation seriously. He ended his fasting after 36 days, losing 33 pounds. Many of his critics lambast Chavez for using the fast as a publicity stunt and say that he has nothing to complain about concerning the pesticides in question. Spokespersons for the grape-growing industry maintained that testing of California grapes for chemical residues showed the samples to be within safety standards. The California Department of Food and Agriculture said that its tests found no residues of the five pesticides protested, except a small level of Captan that was below the 50 parts per million deemed safe by the EPA. Among the most vociferous of Chavez's opponents is the American Council on Science and Health (ACSH). Contrary to what its name suggests, the council is privately funded and, more often than not, functions as a mouthpiece for the industry. With respect to Mr. Chavez's concerns over pesticide usage, the council "condemned what its directors called 'terrorism' and 'junk science' in Cesar Chavez's latest campaign about the alleged dangers to Americans from pesticides." Executive Director Elizabeth Whelan stated, "The reality is that pesticides, when used in the approved regulatory manner, pose no risk to either farm workers or consumers. There is no scientific basis whatever to Mr. Chavez's charge that pesticides cause birth defects, cancer, or any chronic illness. His appeal for funds is based on such hyperbole about risk is unjustified and appalling, particularly when he manufactures his own 'epidemiological' evidence of the hazards of pesticides." According to associate director Edward Remmer, "The Chavez campaign is yet another example of the misuse of science to achieve political ends, which in this case includes an attempt to terrorize uninformed people. ACSH urges consumers to reject this junk science and the appeals of those who distort the scientific realities."

Terrorism, junk science, no scientific basis that pesticides cause birth defects, distortion of *scientific realities?* What does all this mean? Is the "pesticide scare" all just a figment of some vivid or politically motivated imaginations? Not if you ask residents of a new housing development near Bakersfield, California. When a neighboring almond grower flipped a switch on his tractor-drawn tanks and started spraying clouds of pesticide onto his 500 acres of trees, homeowners began to shake and experienced extreme lung pain, which they believe to be caused by the nearby spraying.

Though Richard Wiles of the NAS equates the California situation with "the fox guarding the chicken coop," the state of California is considered to be the toughest in the nation concerning the enforcement of its pesticide laws. One can only wonder what goes on in states with less stringent monitoring agencies. Moreover, the EPA estimates that pesticides have contaminated the groundwater in thirty-eight states, the primary source of drinking water for more than half the country's population. The agency also admits that pesticide residues on food present more of a danger to human health than even hazardous–waste disposal sites or air pollution.

Clearly there are problems with the statements of people such as Whelan and Remmer. There are many who would argue that a toxin is a toxin and that no level of a recognized carcinogen or nerve poison should knowingly be added to our food supply. Even assuming that there could be an established safe level of pesticides in food or the environment, what is there to ensure that these levels are not exceeded? Our monitoring and regulatory agencies, even if given unlimited funding, cannot sample all foods or be at all places where pesticides are applied at all times. Nevertheless, the ACSH continues vociferously to express its convictions on the safety of pesticides.

In the first months of 1989, the National Resources Defense Council helped to bring to the public's attention the risks associated with Alar-treated apples, particularly for children who often consume large quantities of apple products. The American people were furious to learn that the apple, long associated with health, was tainted with an invisible but toxic chemical. So again, the war was on. Since many apple products were used in school cafeterias across the country, school officials were among the first to deny any risk from apple products. "It was overreaction and silliness carried to the point of stupidity," said one school official. Kenneth Kizer, director of the California Department of Health Services said that the Alar scare was creating a. "toxic bogey-man." The ACSH launched an attack—this time by taking out full-page advertisements in *The New York Times*. Again Whelan and other members of the council signed a statement to the effect that: (1) there are no toxic or dangerous pesticides; (2) we would simply starve to death without pesticides; (3) the people who are criticizing pesticide use are doing so without any scientific support—there is virtually no scientific support that pesticides cause cancer or adverse effects; (4) and those against pesticides are groups of alarmists who are actually undermining our health by scaring us away from something that has been safely used for decades.

Commenting on the ACSH's statements, consumer environmental health educator, David Weir, says: "The ACSH is an industry group. It was the chemical industry that sponsored that ad and has a direct interest in promoting the sale of synthetic organic chemicals.

"There is certainly no evidence that we would starve to death without the use of chemical pesticides, although the chemical industry often uses this argument to bolster its position. Chemical pesticides came to us out of the laboratories of World War II. The Nazis developed a whole line of important pesticides, the organophosphates, originally for use in the gas chambers. They were adapted as parathion and malathion to agricultural use after the war, and they have been with us ever since.

"We have used so many chemicals that there is no longer any place to put the residues. Ninety-nine percent of pesticides that are sprayed on crops misses the target and enters the environment as a pollutant. This kind of inefficiency simply cannot continue much longer."

In some of the preceding sections of this chapter we have discussed the manner in which "science" has been used by government agencies to delay action on pesticides. One of the most often used reasons for the EPA's decision not to act on a particular pesticide is that it lacks sufficient scientific data to warrant limiting or banning the pesticide. In and of itself, this reliance on scientific certainty is a cause for alarm. Many critics, for instance, wonder why the government agencies, which are supposed to be protecting public health, do not shift the burden of proof and insist that the chemical industry prove the safety of the chemical. But the scientific issue raises another problem as well. When a government agency either does not test a chemical or allows it to remain on the market because there is "insufficient data" to prove its harmful effects, these declarations start a bandwagon effect among pesticide proponents. Groups like the ACSH, for example, cite this supposed lack of proof as evidence that pesticides are safe.

I recently conducted an interview with John McCarthy, the vice-president and director of scientific and regulatory affairs for the

National Agricultural Chemical Association. His defense of the use of pesticides in today's foods shows how effectively the opinions of "leading scientists" and government officials can be used to deny that the public faces any risk from these chemicals.

"We are concerned about the messages that have been pressed on the public in recent months and the anxiety they create," says McCarthy. "The signals being given to the American public and to the world is that there is something terribly wrong with our food supply. . . .

"But really, the facts are pretty irrefutable that our food supply is safe. . . . This is not only the conclusion of top scientists and officials of the government in the EPA which registers these products and approves their limits on food and the Food and Drug Administration that enforces those limits, but it is the conclusion of leading scientists across the nation. . . . So really, the science is saying one thing, but the public is being given these conflicting messages."

When asked whether he believes that cancer-causing agents should be allowed in our food supply, McCarthy stated that he thought the American public to be adequately protected by the "one in a million" or the "negligible risk" standard established by some of the "leading scientists" in the country.

Safer Choices

Under current pesticide regulation, the EPA must weigh the risks of a pesticide against its economic benefits. Only if this cost-benefit analysis tips on the side of public health can the pesticide be banned or restricted. There are many critics who argue that such a test is inherently flawed from the outset. How can the test possibly be objective? What sort of economic factors can warrant a conscious decision on the part of public officials to ignore demonstrated cancer risks?

This test necessarily assumes that there is some tangible benefit—other than to the pocketbooks of the pesticide manufacturers—arising out of the use of these chemicals. Many times, even once a pesticide is a proven hazard, the EPA will allow continued usage and production until a replacement can be found. But how unique and irreplaceable are these chemicals? Are they really necessary? We grew food prior to World War II without them. So why is it that we suddenly cannot do without them?

Sorely lacking in the EPA's cost-benefit analysis is a consideration of safe and viable alternatives already in use and proving successful. When these alternatives enter into the equation, the economic benefit argument weakens.

Terry Gips is an agricultural economist, the author of *Breaking the Pesticide Habit*, and a former staff assistant to Congressman John Krebs on the House Agricultural Committee. He gives his overview on the problems with pesticides as they are currently used in the United States:

"Since World War II, we have been brainwashed into believing that pesticides are essential in order to produce food. We have set up our whole food and agriculture system to be dependent on them. The result is that we are now in a cycle where we have to use more and more pesticides just to stay in place. That is causing a wide range of health and environmental problems. People really are not aware that there are alternatives to these chemicals."

Fortunately, however, says Gips, we have well-documented scientific research by leading agricultural universities and the USDA that there are alternatives to pesticides.

"In one study," says Gips, "side-by-side farms in the state of Washington were observed. Both were large farms of about 1,000 acres. Both grew wheat. One was a conventional farm, the other an organic farm. The study showed that the organic farm is actually getting higher yields than the farm that is using chemicals and yet it is saving energy, it has less soil erosion and it is more profitable than the chemical farm. It is for that reason that we are seeing an increased interest in farmers who are looking to change from this chemical dependency.

"We are seeing this documented in other studies. In Iowa, for example, a corn grower is discovering that he is saving $75 to $90 an acre in agrichemicals by not using those hazardous pesticides. We are seeing large lettuce growers, major suppliers of the country's iceberg lettuce, who are changing from using pesticides to instead employing a sort of large vacuum cleaner that actually cleans the insects off the crops. So we see the emergence of a number of alternatives to pesticides that are both saving money and not harming our health or our environment."

One thing that the new techniques point to is the fallacy of the claims of pesticide proponents that we need this $22-billion-a-year pesticide industry in order to sustain our agriculture. "Many of these new techniques are based on working with nature instead of working against it, " says Gips. "By working with nature, we can have her do a great deal of the work herself. Part of the problem though is that when you work with nature, the only people who profit are the farmer and the consumer, so many of the industries involved in pesticide production are financially threatened by these new techniques."

Concerning alternatives to pesticide technology, Doug Murray notes that there has been a real lag in research and development on nonchemical substitutes. One of the reasons for this is that "many of the methods do not lend themselves to patents and money-making ventures," he says. Instead, many of the alternatives may be based on traditional or long-lost techniques. Murray gives the example of farmers using a mixture of herbs and plants, or inter-cropping techniques to control pests. But they are told by "agricul-tural technicians" that these are primitive and that they ought to "modernize" their practices. "What we need to see," says Murray, "is a more serious promotion and development of the alterna-tives—on the research and distribution level, both in the United States and in the developing world."

David Weir predicts that by early in the next century the use of pesticides in agriculture will decline dramatically. In fact, he says, we will begin to look back on this type of agriculture as a rather primitive stage, relying on these toxins that kill insects and wildlife along with the insects they are meant to kill.

There are a number of scientifically well-documented, inexpen-sive approaches that control pests. One thing that we know is that some pesticides destroy the immune system of the plant. Just as humans can have difficulty with their immune defenses, pesti-cides in some cases, as well as some synthetic fertilizers, have been shown actually to alter the plant's immune system so that it makes the plant weaker. The plant then becomes even more susceptible to pests, and more pesticides are required to control them.

The idea behind sustainable agriculture is to break that whole pesticide cycle. We try to rely on some old ideas as well as some scientific knowledge that has recently been discovered. Crop rota-tions—changing your crops—and polyculture—not just growing a single crop—helps to minimize pests. There are a number of techniques that can be used. For instance, beneficial insects can be bred and released into the environment to control various pests. For controlling weeds, the Chinese have historically used *weeding geese* that go through the fields eating the weeds, without destroy-ing the plants. The birds leave behind their waste, which helps to fertilize the soil. They also have ducks that will eat the insects.

There is a bacteria that occurs naturally in the environment called *bacillus thuringensis.* This can be formulated into a spray that is nontoxic to people and the environment. A new form of this bacteria, BTI, has been shown 100 percent effective at killing mosquitoes. It can also be used to kill black flies at a 90 to 98 percent efficiency at one seventh the cost of pesticides. Then there are things like India's neem tree. Farmers have traditionally taken the seeds and leaves of this tree and made a pesticide from them that is very effective, but does not harm people. There are many alternatives to pesticides, so that if we only used the synthetics when it was absolutely necessary, I think we would find that we weren't using them much at all.

Often in a discussion of organic farming, those supporting the status quo will argue that organic–farming principles allow the use of naturally occurring, but very dangerous chemicals such as lead, arsenic, and cadmium. This argument is an obvious subterfuge intended to make the public believe that farming is not possible without toxins, so we might as well use synthetic pesticides. Since those farmers practicing the new techniques are interested not only in pest control, but also in producing a healthier food supply; they are not moving from one form of poison to another, but rather to a whole new system that obviates the need for these toxins. Tom Harding, an organic–farming consultant, explains some of the principles of organic farming: "All of the naturally occurring but toxic chemicals such as lead and arsenic are not allowed on the organic farm today. The definition of 'Certified Organic' according to the Organic Food Production Association of North America is 'Certified organic foods are produced, processed, packaged, transported, and stored without the use of synthetic and certain naturally derived pesticides and fertilizers, artificial additives, preservatives, or irradiation. This food is produced with a concern for your health and the environment. It promotes the family farm and rebuilds the soil·through ecologically sound methods and resource stewardship management. It provides high quality, maximum flavor, and nutritional value that is affordable and not cheap."

With organic farming it is not a question of whether we can produce the supply or not, for we can clearly do that. We have to also recognize that even today with our great arsenal of chemicals, we have yet to control the insect. We are losing that war to the extent that pesticide resistance has risen to more than 600 species that are no longer harmed by our chemicals. That makes no economic sense either for the farmer or the farm worker. It also makes no sense environmentally.

There are a number of things that need to be done. We need to reduce large corporate farm projects and encourage family farming and rural community development—keeping people within the farming communities, growing food on a regional basis. We also need incentives to encourage diet, crop, and food diversification. Just because we can live off of four different types of food does not mean that we should not diversify. Very important is the recycling of farm wastes—livestock manures, green manures, and legumes—within the whole cropping system. The development of regional diversified–production systems supporting local farm units and processors can greatly decrease energy consumption. We need to think about buying locally instead of importing everything from California and Mexico.

Skeptics of organic farming also argue that it is not possible to produce food on a large scale using organic techniques. But Harding says many of the farms with which he works are large farms—1,000 to 3,000 acres. "Right now," he says, "we are working with a six–hundred-acre farm growing diversified fruits and vegetables. So, the question is not whether we can do it, but whether we have the will to do it."

Cooperation is essential to the success of any new system. Consumers need to be aware, however, that while a number of innovative farmers are interested in listening to the public, the government and agribusiness may not always be as responsive. We have already seen that our current pesticide legislation does as much, if not more, to protect pesticide manufacturers—to the detriment of the public.

The message to all of us concerned about the safety of our food supply and the state of our environment is that we need to express our views even more convincingly. Politicians need to know that this type of bureaucratic tyranny, particularly when it is at odds with the public welfare, is no longer acceptable. We need to let our leaders know that we want and expect a safe food supply, that we encourage alternatives, and that we want legislation to this effect. We also need to look closely at how each of us can contribute to the suggestions made by Tom Harding—encouraging small farms, regional production, altering our eating habits, and for those of us who can afford it, encourage organic farming by buying its produce. Although these products may cost a bit more at this time, as more farmers see a market, they will enter the market, develop more efficient technologies, and prices will decline. Finally, we need to let not only our politicians and our farmers, but also American industry know that we are tired of being poisoned and that any purposeful introduction of toxic chemicals into our food supply has got to stop—now.

Pollution Solutions

Educate yourself about pesticides, and then educate others about them. A book put out by the International Alliance for Sustainable Agriculture, *Breaking the Pesticide Habit — Alternatives to 12 Hazardous Pesticides* , documents the action you can take to avoid harmful chemicals. Slide shows and a newsletter addressing this information are also available.

A new report has just come out from the NAS talking about alternative agriculture. It explains why we should move government research in this direction and supports farmers moving in this direction. You can request your library to carry publications like this.

Systemic pesticides are inside fruits and vegetables, so they can't be washed away. If you have your own backyard, grow your own food or participate in a community garden program that doesn't use hazardous pesticides. If you can't find organic vegetables, peel and wash as thoroughly as you can. Flowers like marigolds scattered among the vegetables can discourage insects even better than chemicals.

Vote with your dollars by buying certified organic food when you go shopping. Ask food cooperatives, natural, food stores, and supermarkets to carry organically grown food. Vote with your vote. Support candidates who advocate use of pesticide alternatives. Contact leaders at the local, state, and national level concerning legislative initiatives. Letter writing may seem like a small act, but it does give legislators a window on their constituents' opinions. Right now, the International Alliance is trying to incorporate some of the recommendations made by the National Academy of Science into the 1990 farm bill. Organic–standards legislation as well as pesticide reform is being proposed at the national level.

Consider where your money is invested, and try to assure that it be invested in companies that are concerned about the environment. The Sears Coalition, for example, has brought together most of the major environmental groups in the country. The sum total of all members of environmental groups represented across the country is 10 million. We represent over $100 billion in assets that are invested in corporations. If we can use our collective resources to say to companies, "We want you to be environmentally responsible," we can bring about some big changes.

For more information contact:

International Alliance for Sustainable Agriculture
1701 University Avenue, SE
Minneapolis, MN 55414
(612) 331-1099

Chapter 7
Food Additives

In Europe, a traveler can stop in unknown cities, in small villages along the road, and still find fresh, healthy, and wholesome foods. Local farmer's markets are found in almost every town of any size, bread is baked daily, and cheeses and wines are made regionally according to age-old traditions. Traveling in the United States, on the other hand, is a completely different experience. Lining the highways from coast to coast is a steady string of McDonald's, Burger Kings, Pizza Huts, Wendy's, etc.—to the exclusion of all else. We've even looked for truck stops, thinking that there had to be some good old-fashioned home cooking in the heart of America. Somewhere in Kansas, we thought we had found one. At least it had trucks parked out front, and it was not one of the fast–food chains. The experience here was even more disheartening. There was not one fresh thing on the menu—no fruits, no vegetables, no freshly baked bread or biscuits.

Another thing became very apparent as we traveled across the country: the difference between the food revolutions taking place in major urban areas and the way the majority of Americans seemed to be eating. America, on the one hand, is developing some of the most creative and varied cuisines in the world. Many of the restaurants springing up around San Francisco and New York City, for example, could easily compete with some of the best in France or Italy. The "nouvelle cuisine Americaine" uses the highest–quality ingredients, and an abundance of different vegetables and grains, and focuses on variety. This trend toward culinary excellence and creativity goes hand in hand with an increasing population of savvy consumers who look for and demand fresh, diverse foods for their own cooking. In response to a growing consumer awareness about the dangers of food additives and pesticide residues, supermarket chains are beginning to routinely stock a num-

ber of wholesome and nutritious foods—cold–pressed oils, different pastas and grains, organically grown produce. Health–food
stores, particularly in California, are attracting such a large clien-
tele that they can afford to carry a diversity of foods and health–
care products found nowhere else in the world—ranging from
breads, tortillas, and cereals made from organically grown grains,
to sugar-free cookies and jams, to freshly pressed chemical-free
juices, and some are even stocking sulfite-free varieties of wines.

Once you leave the coasts, all evidence of this disappears, and
America once again becomes a vast wasteland of fast, processed,
and fake food. Although the media has much to say about how the
country's eating habits are changing, the fact remains that only a
very small percentage of the population has made much of a
change in the way it eats.

The Philosophy of Food

America always has been a country of extremes and contradic-
tions. This shows up clearly in how we relate to health and diet. On
one hand, we are leading the world in health and fitness, in our
interest in wholesome, nutritious food. An Italian friend said the
other day, "You Americans...You are always *talking* about food. In
Italy, we never talk about food, we just eat it." I thought about that
for a minute and realized it was very true. Good food for Europeans is a way of life. They expect to eat good food—real food—and
consequently they get it. Eating is also a very important part of
European culture. The French take two hours for lunch; in Italy,
the entire nation closes down between noon and 4:00 P.M.. This is
a time for eating, but it is also a time for people to get together,
relax, and give a "specialness" to daily life. Food is what draws
them together.

In America, we have lost that cultural connection to food. People
gobble down meals just to fill their stomachs, many do not even
take a lunch hour or sit down to eat. Most people have no idea, nor
do they care, about what is in their food so long as it tastes good.
For those of us who do care, food begins to take on an importance
equaling, if not surpassing, that placed on it by Europeans. Part of
the reason for this is that good healthy food is so hard to come by in
this country. You really have to look to find restaurants with a
commitment to using only fresh, natural ingredients in their cuisine. If you cook at home, you must be both knowledgeable and
willing to shop around if you want to serve your family just good
basic food.

In many European countries, certain foods are so important culturally that they are under strict control by the government as to their ingredients and their method of production. Parmesan cheese is a good example. Only the *Parmigiano Reggiano* cheese made in the Emilia-Romagna area of Italy can officially be called Italian Parmesan. This staple of the Italian cuisine has a unique flavor and texture that bears little resemblance to products manufactured in this country by companies like Kraft. The French are equally serious about their bread. In France, it is illegal to sell day-old bread in a bakery unless it is labeled as such.

In the United States, however, there are very few guidelines as to how food is to be prepared and sold. In fact, food does not even have to be "real" in order to be called food in this country. And, for the most part, Americans accept what is put before them. Dream Whip is accepted as whipped cream, imitation bacon bits are used on our salads, margarine replaces butter, and fast food is accepted as a valid substitute for home cooking. Cheese not only does not have to be made in any particular way in order to be called "Swiss" or "Parmesan," it also does not necessarily even have to be cheese. In the "ersatz-food pantry" cheese is a favorite with food manufacturers.

Of all the items that food technology has made available, one of the most common is artificial cheese. Producers use it in a variety of prepared concoctions, which are often sold to institutional kitchens. Restaurants use it too, most often for pizza, Mexican food, and salad bars. Though it might not taste quite the same, it usually looks like real cheese, and consumers are rarely aware that it isn't.

The fake cheese is made from powdered casein, the protein portion of milk, mixed with water and vegetable oil, cooked, cooled, and molded. Artificial flavors are added, and if the nutritional value of cheese is desired, vitamins and minerals can also be thrown in. The resulting product is about 25 percent less expensive than real cheese.

Starting on the farm and continuing up to the processing and distribution chains, food production means big business in this country. As with all other large and successful commercial enterprises, the bottom line for food companies is profit. What maximizes profit in the food industry is a uniform product that can be inexpensively produced, and easily transported and stored, with a long shelf life. While fake food fits right into corporate philosophy, why the American public accepts it so unquestioningly is another matter. Why would anyone want to buy an imitation product when they could have the real thing? It is not as if we are in a country where there is not enough grain, fruits, vegetables, dairy, or meat. This country is awash in food. We pay farmers not to produce.

General Foods manager Lois Juliber suggests that convenience is the magic word. The consumer, she says, "wants convenience like she's never had it before." But convenience cannot explain the often radical difference between the real thing and what Americans are willing to accept.

Many people suggest that cost is the determining factor. Food consultant Graham Molitor says that food technology has cut the amount of disposable income Americans spend on food from 22 percent in 1945 to 15 percent in 1977. Others say that the interest in healthier, fresher food is merely another yuppie fad and that only an elite few can afford to eat these foods. Again the question arises, if indeed this is the case, why is it in this land of abundance that only the wealthy can be well-nourished? Admittedly, fake food is cheaper to produce than the real thing. So are high-tech real foods. For example, among the many drugs given to beef cattle are the hormones testosterone or progesterone. The drugs can cut down by 21 days the time it takes for an animal to reach 1,000 pounds, and yields leaner meat at the same time. These drugs then reduce the cost of producing beef to the rancher, which eventually is reflected in lower meat costs to the consumer. But are these *real* cost savings? And, more important, are they worth the health risks associated with hormone-treated meat? The Europeans think not. In 1985, they banned the use of such drugs in their own cattle, and in January 1989 banned U.S. imports of hormone-treated beef.

Cheap food is not really any less expensive when you calculate all the associated costs. Pesticides not only leave residues on our fruits and vegetables, agricultural runoff pollutes our waters, making much of our seafood inedible. Pesticides are not inexpensive, and even with their extensive use, one-third of the country's crops are lost to pests. When all of these costs are calculated, food produced with pesticides hardly looks either efficient or cost-effective.

Moreover, while many commercial foods carry a low price tag, others do not. Soft drinks are prime examples. Americans drink an average of 400 twelve-ounce cans each year. Coca-Cola is advertised in more than 80 different languages around the globe. More than 200 million people throughout the world consume some 12.5 million gallons of Coke each day. In 1988, Pepsico had gross profits of about $13 billion, and netted around $760 million. This is money spent on a nutritionally valueless product. Both have targeted China as the new frontier for soft–drink marketing. Starved for Western culture, the Chinese people have welcomed the soft drinks. The Chinese press and government, however, see things differently. *The New Observer*, a monthly Beijing publication, refers to Coke as "a useless luxury that is harmful to people and the state." The magazine added that the sugar and caffeine in Coke is both unhealthy and an insult to Chinese national pride.

Advertising is probably the single most important factor influencing what Americans eat. The key to advertising is illusion, making things appear to be something other than what they really are. Sometimes this is done through focusing on what is believed to be an attribute of a particular product. Milk advertisers, for instance, concentrate on protein and calcium, and tell us that "milk helps build strong bodies." Young, fit, and healthy teenagers usually star in these advertisements. What we are not told is that homogenized and pasteurized milk promotes cholesterol buildup in the arteries, that milk is poorly digested by large numbers of people and may even contribute to calcium deficiencies. Other advertisements are outright misleading—beer, soft–drink, and fast–food ads being the worst offenders.

While the United States may lead the world in the promotion, manufacture, and consumption of fake food, the public's response has also been typically American. Not only do those people interested in health and fitness eat better and exercise more than their counterparts in other nations, but food has been tackled head-on by forceful and determined consumer advocacy–groups. As a result of pressure from these groups, America now has the most stringent food–labeling laws in the world. Under these laws, ice cream cannot be called "ice cream" unless it really is; fruit juice can no longer be called "juice" unless it contains some of the real stuff; and all products must list ingredients in descending order according to their proportions.

Food labeling does not make up for adulteration, but it does, to a certain extent, put the ball back in the consumers' court. Consumers now have the option of knowing what goes into their foods and in what concentration. Before the era of food labeling, for example, most of the commercial cereals had sugar as their number–one or number–two ingredient. With labeling and the public's awareness of the health risks in sugar, there are now quite a few commercial cereals on the supermarket shelf that contain little or no sugar. Still, in order for labeling to be really effective, consumers have to be willing to assume a certain responsibility. First, people have to take the time to read the labels. Once you start to do this, you will notice that there are alternatives to products full of chemicals. If you buy canned tomatoes during the winter to make sauces, you'll see that 9 cans out of 10 will have the same ingredients: tomatoes, tomato juice, and calcium chloride—a chemical added to preserve color and texture. If you look long enough, though, you'll usually find a can that does not contain calcium chloride. The tomatoes are just as red, just as firm, and just as good for your sauces.

Unless you are both alert and diligent, however, this still may not be enough. During the past 10 years or so, many supermarkets have begun to stock foods like whole-grained breads and cereals. They carry labels saying *Whole Wheat, All Natural, Contains Oat Bran*. When you read the fine print on the labels, you will see things that do not sound quite as healthy—emulsifiers, dough conditioners, monoglycerides and diglycerides, and, unless the products specifically says *No Preservatives*, you will probably find things like calcium propionate, BHA, or BHT. This can be confusing, especially since almost all commercial bread products contain these ingredients. But this is also the point at which labeling has the potential to become particularly effective. You may decide that even though you do not understand what all these things are, you simply do not want them in your food any longer. Taking bread as an example again, the only ingredients that are necessary to make good bread are flour, water, yeast or starter dough, and depending upon your tastes, salt and sweetener. Bread dough does not need to be "emulsified" or "conditioned"—unless it is not kneaded properly. Nor does it need preservatives—unless it is being sold stale. Consequently, you may decide that you need to go elsewhere to get the type of product you want. It is you and others like you saying "No" to ersatz food and going out of your way to make sure that you get the real thing that will cause changes within the food industry.

Health and Diet

Notwithstanding the claims of our doctors and health professionals that Americans are living longer than ever before, you only have to look around you to see that, as a nation, we have never been more unhealthy. A quarter of a million Americans are diagnosed with cancer each year. In 1988, federal researchers prepared a report concluding that notwithstanding greater awareness about the benefits of exercise and a healthy diet, many Americans remain "significantly overweight." We are a nation of obese people—one quarter of the entire population is more than 20 percent above their proper weight. We suffer from some of the highest rates of heart disease, osteoporosis, arthritis, and high blood pressure in the world. It is no coincidence that most Americans' diets consist mainly of red meat, fatty and fried, sugary, high-calorie, low-nutrition foods.

The American medical establishment has been slow to recognize the health-diet connection. This is changing, but only in a piece-meal manner. Our doctors now tell us to avoid cholesterol if we want to decrease the risk of heart attacks, high blood pressure, and atherosclerosis. We should eat more fiber if we want to avoid colon, prostate, and stomach cancers. But how helpful is this kind of advice to the average American? We submit that it is very helpful to the American food industry, but does little to help the average consumer.

Very little is *really* known about the human body. It is a complex of interacting and interrelated systems. Modern medicine and science by their analytical nature tend to view health and the body as isolated systems—the heart, the lungs, the brain, etc. Their advice about health is thus specific. For instance, atherosclerosis is caused by an accumulation of plaque in the arterial walls. As these deposits accumulate, the openings of the arteries grow smaller. They can eventually constrict to the point of causing a heart attack or stroke. So, says the medical establishment, if you want to decrease your risk of suffering a heart attack or stroke, you should decrease your intake of cholesterol, since it is the cause of the plaque accumulations in your arteries. A specific solution to a specific problem.

But the body does not work like that. In the case of cholesterol, most medical authorities today believe that you increase the level of the fatty substance in your blood by eating foods containing it, such as meat, cream, butter, and eggs, but this is not at all a scientific certainty. Cholesterol is naturally manufactured within the body, primarily by the liver. It is essential for digestion and the production of sex hormones. Cholesterol may be synthesized within the body from foods that contain no cholesterol themselves. Consequently, there is ample reason to believe that either these foods stimulating cholesterol production, or a faulty metabolism of the substance, play at least as important a role in raising serum cholesterol as the intake of high-cholesterol foods. In fact, research conducted at the Stanford University School of Medicine indicates that drinking more than two cups of coffee per day increased blood cholesterol levels in men. Approximately 120 million Americans drink more than three cups of coffee a day.

Nevertheless, cholesterol is the new medical buzzword, and it has been taken up with enthusiasm by the food industry. While the industry could not have withstood an attack by the medical establishment on the general quality of its products, food manufacturers have had little trouble adapting to lowering cholesterol. Beef producers are producing leaner beef (through hormones), margarine ads abound like never before, and poultry sales have skyrocketed. Dairy farmers are undeniably suffering from the bad press, but even they are not total losers.

Ironically, in promoting these more "healthful" products, the American consumer may be getting even less nutritious food. Butter is a high–cholesterol food, so are cream and cheese. But these foods are good sources of vitamins A and D. Compared with many other commercial foods, they are also among the less adulterated. Many food manufacturers have made a good profit in selling substitutes for these foods—margarine, artificial whipped cream, nondairy coffee creamer. These products are filled with chemicals—emulsifiers, binders, colors, and preservatives. Additionally, many of these dairy substitutes are made from coconut or palm oil, both of which are higher in saturated fats than cream, lard, butter, or beef fat. In a survey of 25 nondairy coffee creamers conducted by two University of Nebraska medical professors, they found 22 contained coconut oil. Another study, conducted by professors at Texas A&M, showed that when healthy young men were fed a diet containing about 20 percent of the calories from coconut oil, their blood cholesterol rose significantly. A diet with 20 percent meat fat, the professors found, had little effect.

Margarine is made either in whole or in part with hydrogenated vegetable oil. Once an oil is hydrogenated, it becomes a saturated fat known to increase cholesterol levels. Consequently, even though a margarine is made with sunflower, safflower, or corn oil—oils that are naturally low in saturated fats—the end product is a food with high saturated fats.

Cutting cholesterol alone is hardly an assurance of total health, particularly when it is done by merely switching from a real food to a fake one that is high in chemical additives and saturated fats. The American medical establishment is adopting a somewhat more holistic or integrated approach to health when it recommends that Americans cut down on their intake of red meat, and increase their intake of unrefined grains, fruits, and vegetables. Such changes would require major dietary and perhaps even lifestyle shifts for most people in the United States. Not difficult changes, but changes many people are unwilling to make for themselves. Nor could the commercial food industry deal with any great upheavals in America's eating habits. Consequently, what we have seen as the practical result of these recommendations is the emergence of another buzzword—fiber.

Fiber is a concept that the general public can grasp, and is one that is easily marketed by the food industry. It is true that eating less meat, more whole grains, and more fresh foods will increase fiber in the diet. Also true is that the typical American diet is very low in fiber and that many of today's most common ailments—constipation, obesity, diverticulitis, andcolon and prostate cancer—are directly related to inadequate fiber in the diet. But adding a tablespoon or two of oat bran or a bowl of high-fiber cereal to a fundamentally deficient diet is not going to get at the root of these health problems.

Anywhere from 25 to 50 percent of the average diet in this country is composed of animal proteins such as meat, poultry, milk, cheese, and eggs. All of these high-protein animal products are virtually devoid of fiber. While adequate protein is essential to build, maintain, and repair all parts of the body, excessive protein can be harmful. The liver is the body's major detoxifying organ. It is also the organ responsible for the production of bile, a substance used for protein metabolism. Excessive protein not only taxes and weakens the liver, but can also result in overproduction of bile, which can lead to diseases like gout and arthritis. While fiber will work to mitigate these conditions, the only truly effective way to combat them is to decrease protein consumption.

It is no coincidence that all of a sudden people are talking so much about fiber. Prior to the advent of fast and processed foods, most people obtained adequate fiber from the foods they normally ate. Whole-grain bread, for instance, is an excellent source of fiber. The only reason that fiber must be added back into the bread today is that it is removed during processing. But is it enough to just add something back once it is taken away? Food processors are fond of saying that there is no nutritional difference between processed bread and whole-grain bread because they have added back all the vitamins and nutrients destroyed by processing. This is both untrue and misleading. Certain nutrients can be supplemented fairly easily, fiber being one. Fiber itself is a nutritionally neutral substance. Its role is not to be digested, but to add bulk to what we eat so that food is moved easily through the intestines. But vitamins and other nutrients are another situation. Most vitamins occur naturally in foods in forms and concentrations that are easily metabolized by the human body. The B vitamins, for instance, have long been recognized as occurring in a complex. Most commercial vitamin-supplementing only accounts for a few of these elements—thiamine (B-1), riboflavin (B-2), pyridoxine (B-6), niacin (B-3), and sometimes vitamin B-12. Elements such as folic acid, choline, and pantothenic acid, although important for total health, are removed by processing but rarely supplemented. Nor does supplementing respect the natural concentrations of vitamins that aid in their absorption.

Vitamin C is another nutrient that appears naturally as a complex. This vitamin is extremely sensitive to heat and other forms of processing. Orange juice and citrus–drink manufacturers claim that their products are just as nutritious as fresh–squeezed juice because they add "vitamin C." This is also untrue. What is added back is usually ascorbic acid, or sometimes sodium ascorbate (a buffered form), but other elements of the complex are not replaced. Some elements of naturally occurring vitamin C are essential for collagen production, a gluelike substance that holds our skin, bones, and tissues together. They also aid in the proper assimilation of the ascorbic–acid component of the complex. Vitamin E, commonly found in unrefined grains and essential for the production of many of the body's hormones, is likewise destroyed by processing. It is rarely supplemented at all.

Food Allergies and Diet—How They Affect the Way We Act and Think

Most people are familiar with allergies. They are those uncomfortable sneezing attacks you may get from pollen and ragweed, or that itching of the eyes you experience whenever you are around the pet cat or dog at a certain friend's house. Some people may get rashes from an allergic reaction to strawberries or other foods, while others react so violently to a particular food they need to be rushed to the hospital if ever they inadvertently ingest it.

But Oklahoma psychiatrist William Philpott has long studied the effects of environmental factors and foods in causing violent behavior. In one case, a 12-year-old boy became so violent after eating a banana that he tried to hit another child with a stick. Eating an apple, the same child started a fight. A 52-year-old woman, when tested for wheat allergy, said that she felt like hitting someone. In another case, an 18-year-old boy became delirious when tested with tobacco, hit the examiner, and began hallucinating that the examiner had horns on his head and was the devil.

These findings of food-related allergies and behavioral problems come at a time when the American diet has never been more adulterated and nutritionally deficient. Is this a mere coincidence? There is ample reason to believe that it is not. For a number of years, doctors have recognized the effects that dyes and food colorings can have on children; hyperactivity, bed-wetting, irritability, and learning disabilities. All of these conditions are precursors to the types of violent behavior that later can cause teenagers and adults to become criminals. Additionally, many clinical ecologists (doctors and allergists focusing on the link between behavior and food) are discovering that the foods we consume most frequently are among the worst offenders. Meat, dairy products, corn, citrus, and wheat are particularly troublesome for many Americans. One reason that Americans may be suffering as much as they are from allergies today—both traditional and cerebral—is that our diets include an over consumption of too few foods. A typical diet in this country consists of about 25 percent dairy products, 25 percent meat and poultry, and another 25 percent wheat products. The remaining portion is made up of sugar, fat, and a small portion of fruits and vegetables. Even if you don't eat corn-on-the-cob or other products obviously rich in the grain, if you purchase commercial foods, there will be a large amount of corn and corn products in your diet. Be it corn oil, margarine, or cornstarch, corn is virtually ubiquitous in our food supply. Chicken and cattle are also fed with it, so corn residues can be found in meat as well.

Another reason that commonly consumed foods may be causing an increased incidence of health and behavioral problems has to do with the chemical treatment and processing of these foods. I have had experience with this firsthand. For a number of years, I believed myself to be allergic to orange juice. Whenever I had a glass, I came down with classic allergic symptoms: itchy eyes, runny nose, and sometimes migraine headaches. On one occasion, I was visiting a friend who owned an organic farm in the country. For breakfast, he served fresh-squeezed orange juice. It smelled and looked so delicious that I drank a glass. I waited and waited, but no symptoms occurred. Then, a few months later in Florida, I had a glass of juice and all my symptoms reappeared. I realized that it was not the juice, but something in the juice, or perhaps in the fruit itself—most likely pesticide residues or the chemical changes that took place when the prepackaged juice was pasteurized.

The way America produces and processes its food may very likely be having similar effects on large numbers of people. Take milk, for example. People throughout the world have been using dairy products for thousands of years. It is aged to make cheese, the cream is skimmed off to make butter, and it is fermented to make yogurt or kefir. Traditionally, these products have been made locally and on a small scale. For the past forty years in America, dairy production has left the small farm and become the domain of big business. In order to produce more milk, dairy farmers today routinely administer hormones; to cut down on disease, animals are sprayed with pesticides and antibiotics; cattle feed is laced with pesticides. The residues from all of these chemicals are present in the milk and dairy products you buy.

Milk contains many enzymes that are essential for the proper digestion of milk products. Heating destroys these enzymes, and as a consequence, undigested molecules of milk may enter the bloodstream. The immune system recognizes these molecules as foreign and attacks them, provoking an allergic reaction. All commercial milk today is pasteurized, a process by which the milk is partially sterilized by heat treatment. Pasteurization was a great discovery in the days before refrigeration and proper sterilization procedures were known. Today, its primary benefit is to save dairy farmers the expense of having to institute proper sanitary conditions, conditions that may be altogether unattainable in a large-scale mechanized dairy setting. Pasteurization is also responsible for the destruction of vitamins. Vitamin B-6, for example, can be reduced by as much as 50 percent.

Homogenization is a process in which fat globules in the milk are broken apart so that milk has a more uniform texture. It is the reason that our milk no longer has that layer of cream that used to rise to the top of the bottle. Homogenization reduces the size of the fat globules, making it possible for these molecules to be absorbed by the small intestine. This not only increases the potential for allergic reaction mentioned above, but also may be a major cause in itself for atherosclerosis.

Pasteurization and other processing techniques also affect the quality of cheese in this country. Making cheese is an ancient art. Flavors and textures vary from region to region, and from country to country depending on the grasses used for grazing, the traditions behind manufacturing techniques, and the types of milk used. Aging has always been an essential part of making cheese. In America, cheese is something altogether different. Here we can make cheese in a matter of a few hours or days, and, no matter what it is called, it looks and tastes the same. Rennet is used by commercial manufacturers to produce cheese quickly and in large quantities. This substance is derived from the fourth stomach of young calfs and is so powerful that 1 ounce can clot 2.8 million quarts of milk. Pasteurization destroys the enzymes and bacteria essential to fermentation and giving each cheese its distinctive flavor. Commercially produced cheese can contain a large number of chemicals that may or may not be listed on their packaging labels. Stabilizers and mold inhibitors are used to retard spoilage and provide a long shelf life. Colorings and bleaching agents are used to regulate appearance. Benzoyl peroxide is a commonly used bleach in cheese manufacturing. This chemical is also used as a hardening agent in fiberglass and is believed to cause skin allergies.

All of these additives and processing techniques make American dairy products a very different food from those traditionally consumed throughout the world. At the same time, a growing number of people in this country are finding themselves allergic or sensitive to milk products. This seems to be a distinctly American phenomenon. We have had a number of European friends who have remarked upon this, saying, "We just do not understand why you Americans are all allergic to milk. In my country, I've never heard of someone with that kind of allergy, and we have been eating dairy products for many, many years."

<u>Fake, Denatured, or Adulterated?</u>

All commercial food in the United States has been altered, adulterated, and tampered with in some way. Raw foods, fruits, vegetables, and grains are all sprayed with pesticides. Meat and dairy animals are sprayed with pesticides, shot full of hormones, and fed grain containing still more pesticides. Before any of these foods reach the consumer, they undergo additional treatments. Grains, for instance, are degermed, bleached, and usually preserved with a chemical. Meat is treated with aging and flavoring chemicals, nitrates are added to preserve color. Consumer advocate Ralph Nader has pointed out that making food appear what it is not is an integral part of the $125 billion food industry. The deception ranges from the surface packaging to the integrity of the food products' quality to the very shaping of food tastes. The industry's catering is calculated to sharpen and superficially meet consumer tastes at the cost of other critical consumer needs. These tastes include palatability, tenderness, visual presentability, and convenience.

Because high-cholesterol foods are mainly "real" foods like meat and dairy products, the cholesterol scare has been a challenge for the food technicians who are scrambling madly to discover innovative alternatives. So far we have received things like margarine, bacon bits, and nondairy coffee creamers. Technicians at Proctor & Gamble, however, have set their sights even higher. The Cincinnati-based company whose reputation and fortune has been made by turning fats into consumer necessaries like soaps, detergents, Crisco, and Oil of Olay, has spent the past 20 years developing the ultimate consumer product—a zero-calorie fat substitute. The new product, sucrose polyester, called Olestra by the company, is now the subject of a 10,000-page petition for licensing before the FDA.

Olestra is a fat substitute. The company has asked the FDA to approve its use in shortenings and oils and in the preparation of certain fried snacks like potato chips. The thing about Olestra that makes it special is that is has no calories whatsoever, it's digestively inert like fiber. It is, as Michael W. Tafuri, director of Procter & Gamble's Olestra product–development division describes it, "zero nutritious."

Not only are American food producers pros at creating and marketing zero-nutritious foods, but they also excel when it comes to "negative nutrition." Top on the list in this category is the $50-billion-a-year fast-food industry. According to statistics from the National Restaurant Association, 45.8 million people, or about one fifth of the population—eat at fast-food restaurants on an average day. In 1987, McDonald's alone had close to 10,000 outlets worldwide. Ninety-five percent of these were in the United States.

Despite the friendly, healthy images the fast-food conglomerates sport on television ads, the reality remains that the food they serve is eroding health nationwide and worldwide. What ads conveniently overlook is that fast food is high in fat, sodium, sugar, and overall calories.

Even though diseases like atherosclerosis and heart attacks are associated with older people, scientists are finding that first stages of plaque buildup in the arteries can begin at age two. Speaking about the effect of fast food on children, Tazewell Banks, director of the Heart Station at D.C. General Hospital in Washington, says, "We're clogging up their arteries. If you hate your children, you tell them to play in the street. Not quite as bad but in the same direction, if you hate your children, send them to those fast-food places."

Under public pressure, McDonald's shifted away from its practice of using beef tallow for its fried foods. While this reduces the risk of fat-related disease, there still remain many things wrong with fast food. Researchers and psychologists are also finding that junk food may be responsible for a host of behavioral problems in children. Hyperactivity and aggressivity have been linked to "overconsumption-undernutrition" syndrome. These children eat too many calories in relation to the vitamins and minerals they derive from food. This disturbs the brain's chemical balance and its metabolism of nutrients. When candy and ice cream are replaced with fruits, vegetables, and complex carbohydrates in conjunction with B-12 supplements, behavior improves dramatically.

Food and Politics

No matter what the issue in America, the way it is fought out in the political arena is always the same. Food is no exception, and politics is an important factor in making American food what it is. The scenario looks something like this. A consumer group or an association of concerned citizens will become aware of a problem. It does not matter what the problem is. It could be the pesticides DDT, Alar, or Temik, or it could be hazardous–waste disposal, nuclear–power plants, or food irradiation. Whatever the problem, the reaction is always the same. The industry affected will band together with its public–relations firms and its lobbyists and denounce the problem as a fabrication. It will defend its product as totally safe, tell the public that it is overly susceptible to the hysteria of a few fanatics. The level of governmental involvement often depends on the pushing power of the citizen groups involved.

Concerning the safety of our food supply, one substance after another has been involved in this process.

Dyes

Synthetic food coloring plays a major role in today's food processing. It can be used to make butter as yellow as daffodils, make meat appear red and juicy, and is often injected into fruits and vegetables to make them appear fresh and vine-ripened. The United States leads the world in the consumption of synthetic dyes. The need for these coloring agents increases every year as more and more foods become ready-made, canned, bottled, and are shipped great distances, then stored and stocked. In this way stale and unappetizing food is given a face–lift, and few buyers know what the real substance actually looks like.

The FDA is the federal agency responsible for the licensing and approval of dyes for use in foods. Certification of dyes began in 1938 with the passage of the Food, Drug and Cosmetic Act. At that time, 19 dyes were certified. By statutory definition, certification of a dye means that it is pure and nonharmful no matter how much of it is used. No legal action may be instituted against a food manufacturer for using even an excessive amount of coloring so long as it is certified.

These artificial food colorings were in wide use for almost 20 years before laboratory tests revealed what should have been apparent to the FDA when it granted certification, namely that the majority of the dyes were carcinogens. Under substantial public pressure, the FDA decertified 9 out or the original 19. An additive alert has been issued by the Center for Science in the Public Interest for three dyes still in common use: Citrus Red #2, Red #40 and Yellow #5.

The FDA is not the only federal agency whose performance in regulating artificial colorings has raised substantial questions of ethics in government. In June 1985, a House committee accused the Department of Health and Human Services of illegally allowing the continued use of six known cancer-causing dyes used in a wide variety of products. The congressional report found that despite findings by the FDA as to the carcinogenicity of the dyes in question, HHS failed to act to take the dyes off the market. "This report," said Congressman Ted Weiss, Democrat from New York, "reflects Congress's bipartisan view that HHS has blatantly violated federal law in failing to ban six potentially hazardous dyes."

The best known of the dyes in question was Red #3, widely used to color maraschino cherries, alcoholic beverages, and fruit cocktails. According to a Capitol Hill source familiar with the controversy, officials at both HHS and the Office of Management and Budget were heavily lobbied by maraschino cherry producers, as well as drug and cosmetic manufacturers, to keep the coloring on the market. Industry pressure, he said, explained the reversals in the government's position concerning the dyes. FDA officials and a former assistant health secretary had issued recommendations to ban some or all of the chemicals. A year prior to the House investigation, FDA commissioner Frank Young told Congress that the FDA had completed its studies on the dyes. A year later, just before the House report was issued, Commissioner Young ordered new evaluations of the studies. This review, said the House report, is "patently unnecessary" and in effect will "subvert, if not completely paralyze, enforcement of the nation's public health and safety laws."

Four years later, these predictions of subversion and paralysis proved true. In July 1989, the House approved a study on Red #3, the only food coloring on the market whose safety is still being debated.

The Food and Drug Administration had planned to ban some uses of the dye in the following months. But it said it would not take the step if a bill was signed into law authorizing a new study of the dye. After the study is completed, which could take three years, the agency would make a decision on banning the dye. And the FDA, still headed by Commissioner Young, is unable to determine whether the carcinogenic effect of the dye was caused directly by Red #3 or whether it should be considered a less dangerous secondary carcinogen, creating conditions that increase susceptibility to cancer.

Given the congressional pronouncement three years earlier that additional review was "patently unnecessary," the House's bill mandating such a review three years later is ambiguous to say the least. It should come as no surprise however, that this dilatory legislation was supported by the pear and cherry industries, both large users of Red #3.

Cyclamates and Artificial Sweeteners

The FDA's reluctance to interfere with a free market in foodstuffs produced by the nation's large corporations is also illustrated by the agency's approval in 1950 of a pharmaceutical company's new artificial sweetener, a cyclamate product. Consumer advocate James S. Turner reports that if decisions had been made on the basis of the company's research data, the manufacturer's request to market cyclamates would have been rejected. Instead, the FDA took the unusual step of running its own laboratory research on the product and used this data as the basis for its approval.

Shocking as it may seem, eve the FDA research clearly noted the appearance of malignant tumors in test animals. Nevertheless, this manufacturer and other companies subsequently developed cyclamates that were approved without restriction by the agency. According to Turner and other consumer advocates, warnings about the dangers of cyclamates were given to the FDA from the mid-fifties through the late sixties. All of these alerts were ignored. It was not until October 18, 1969, that the then-secretary of health, education and welfare Robert H. Finch announced the prohibition of cyclamates. Meanwhile, for almost 20 years, the FDA had allowed the widespread usage of a product its own research had identified as carcinogenic.

This type of action (or inaction) is common when the FDA is dealing with large corporate interests. The agency's approach to small businesses or individuals is quite a different matter. As long ago as 1965, Senator Edward Long of Missouri conducted hearings that he said, "uncovered instance after instance of FDA raids on small vitamin and food supplement manufacturers. These small defenseless businesses were guilty of producing products which FDA officials claimed were unnecessary for the average human diet." At the same time, the FDA was approving additives and processing techniques that not are not only "unnecessary for the average human diet," but are actually harmful and capable of causing cancer.

The failure of government to adequately protect the public interest can be largely explained by a revolving–door policy. This is a closed system where conflicts of interest abound, starting with the "donations" elected officials receive from industry. Once in office, the elected official has certain political appointments he or she can make. In the case of the president, these appointments will be the heads of agencies such as the EPA, the FDA, or Health and Human Services. More often than not, these political appointees will either be drawn from the industry the particular agency is mandated to oversee, or will be approved by the industry as supporting their position. Upon retirement from the agency, it is not uncommon for the government official to go to work for the very industry he or she has been charged with regulating.

The American Council on Science and Health

In the usual course of politics in America there are also a number of other participants giving support to the industry position. Among the most influential and publicly visible of them is the American Council on Science and Health (ACSH), a group representing itself as an objective and unbiased scientific organization. Contrary to what its name suggests, this group is not a consumer education and advocacy group, as it is commonly referred to by the media.

The ACSH has issued numerous reports since its inception in 1978 on various topics including dioxin, EDB, formaldehyde, and vitamin B-15. Typically, when the ACSH issues a report or a position statement, the conclusions of the report are reprinted in a shorter summary version, which is widely distributed and available to the general public. Another common practice is for the abridged version to conflict dramatically with the scientifically based conclusions found in the original full-length text. Since the public and the media rarely take the time to research background material, these conclusions are usually accepted as unbiased scientific facts. But, says Peter Harnik, a journalist for the Center for Science in the Public Interest, the council's omissions, representations, and factual errors always seem to be in the direction that the group wants to take, and opinion seems to be substituted for scientific fact when reality does not match the group's needs.

Just what are the needs of the ACSH? Why is it consistently one of the most outspoken opponents of any regulation or limitation on the use of chemicals? What is the purpose of always insisting that there is not enough "scientific" proof that a chemical is dangerous? We now know that once a chemical becomes suspect of causing damage, more often than not, the suspicion is well founded. So, if the ACSH is truly an unbiased "scientific" organization and "consumer" group, why do its positions invariably support industry and disregard the public interest?

According to Ralph Nader, a consumer group is an organization that advocates the interest of unrepresented consumers. It must either maintain its own intellectual independence or be directly accountable to its membership. In contrast, the ACSH is a consumer-front organization for its business backers. It has seized the language and style of the existing consumer organizations, but its real purpose is to glove the hands that feed it.

ACSH reports typically declare that the substance under review does not pose any significant health risk to the American. They also invariably argue against imposing any limitation or restriction on its use as being unwarranted and premature, usually maintaining that more scientific study is necessary before any action should be taken. The value of the ACSH's veneer of objectivity is in its ability to appear credible and thus draw significant media coverage when it issues statements on controversial subjects.

Among the examples of this newest form of industry politicking are:

•**Washington Forest Protection Association**: A trade group made up of the largest timber companies in the state of Washington. The group fights legislation concerning logging restrictions.

•**Clean Air Working Group**: Composed of representatives of the oil, steel, aluminum, paper, and automobile industries who lobby against tougher amendments to the Clean Air Act.

•**The National Environmental Development Association**: A group of industry representatives whose position is that "the financial burden of unjustified environmental measures could destroy industry's ability to provide the tax base for general social progress."

Putting the Pieces Together

So far we have discussed a political process that gets under way with a consumer group bringing an issue before the public. The typical response of the government agency involved in regulating the matter is either back up the industry or deny that a problem exists. In the debate about issues concerning diet, health, and nutrition, the medical and university establishment also comes into the picture to support the status quo. Commonly any complaint raised by consumers about food or nutrition will be denied by these groups, who insist that there is absolutely nothing wrong with our food supply. The role of the media in most debates concerning health and the environment is to parrot what is said by the government agencies, and thus to also support the industry position.

The Food and Drug Administration

The FDA is the federal agency responsible for the oversight of many aspects of our food supply. It is the FDA's job to inspect fruits and vegetables for pesticide residues and to monitor meat and poultry for disease. The FDA is also the agency responsible for licensing chemical additives, food colorings, and certain food–processing techniques. Although the FDA's primary raison d'etre is to protect the public from harmful substances within the food supply, this agency is unerringly pro-industry. Despite widespread public opposition, the FDA, upon its own initiative, approved food irradiation. Many scientists believe that the agency acted in spite of strong evidence indicating that irradiation is both carcinogenic and mutagenic.

Eating Naturally

Food is one of the most talked–about subjects in America. Diet books are number–one best–sellers and rate highly among favorite topics with talk-show hosts. Food is one of the most advertised products on television—from Saturday morning cereal ads target at children, to images of fitness and glamour used to advertise soft drinks and fast food. Network news programs discuss the latest medical findings on cholesterol and fiber. Yet with all this discussion, very little changes— more Americans suffer from heart disease, obesity, and a range of other diet-related health problems.

Eileen Kugler, Director for Communications at The Public Voice for Food and Health Safety, offered us some ideas on food safety. She suggests we support farmers' markets and try to buy fresh fruits and vegetables locally. Small farmers are making an attempt to reduce the toxic residues on their produce. Also shop at co-ops and natural food stores.

Find out what supermarkets are doing to provide alternatives. Do they have pesticide–residue programs of their own? What percentage of the foods they offer are actually tested? Do they subscribe to any service that provides them fruits from organic sources? Show them that there is a market for organic produce by requesting it.

Many manufacturers and food packagers have 800 numbers listed on the products. Call them and ask questions. Also let them know that you want information on the labels of their products.

Wash fresh fruits and vegetables and peel them when possible. Don't let any meat or poultry sit out at room temperature for more than two hours. Wash poultry very carefully and be careful not to cross–contaminate a cooked product with a raw one. For example, after you cut up a chicken, carefully wash your cutting board and hands in warm, sudsy water (not just a rinse) before you cut up salad or anything else you're not going to cook on the same board.

Buy seafood only from a reputable source, from an established seafood market or restaurant. Ask where the fish comes from. Be sure to eat a variety of foods. That way you have less potential for a food that builds up a particular toxin.

Buying fruits and vegetables locally and in season is the best way to avoid those that have been treated with fungicides as they were being transported.

For more information contact:

Public Voice for Food and Health Policy
1001 Connecticut Avenue, NW , Suite 522
Washington, DC 20036

Chapter 8
Food Irradiation

The United States Department of Energy (DOE) has a problem. Its plutonium–manufacturing plants are falling into disrepair, and nobody wants to spend the billions required to bring them up to snuff. Consequently, the nation's capacity to keep up a large-scale, sophisticated arms race is seriously impaired. There is an even bigger, related problem. Over the years, making bombs in America has been a very dirty business. An environmentally aware public is beginning to see just how dirty. Some estimates on the cost of decommissioning the nation's aging nuclear–weapon complex will be in the hundreds of billions of dollars. Those taking a realistic view of nuclear–waste disposal—from military as well as civilian sites—estimate that there is not enough money in the Gross National Product (GNP) of the entire planet to effectuate a total cleanup of the nuclear mess.

The DOE has proposed waste sites throughout the country. While the public is adamant about the department safely disposing of its radioactive paraphernalia, no one seems to want a nuclear–waste repository in his backyard. What can be done?

For a number of years, it appeared that the government did have a solution to all of its problems—it could dispose of the waste, come up with new sources of plutonium, and probably even make money in the venture. The only hitch was to get the public to buy the idea. Enter the concept of food irradiation. It is a processing technique performed by exposing food to ionizing radiation. It is designed to prolong shelf life and, in some instances kill harmful bacteria. There are currently three techniques being considered for the process. Either the radioactive isotopes cobalt-60 or cesium-137 can be used, or the food can be exposed through a large electron beam that resembles an oversized X-ray machine. Everything from milk to canned vegetables to rice and flour can be irradiated to be preserved.

With the American public, press, and even the Congress criticizing the DOE for inefficiency and ineptitude in managing its nuclear–waste disposal, the ideal solution would be to recycle it. The magnitude of the DOE's waste problems requires generating a virtually unlimited demand for its recycled by–roducts. Everyone in the United States, or the world for that matter, has to eat, so food is an ideal market, with the population of the world generating all the demand that the DOE.could ask for. Furthermore, with people concerned about pesticides and chemical contamination, it would be easy to promote food irradiation as a wholesome, nonchemical solution in much the same manner that nuclear energy was being sold as the answer to environmental concerns about greenhouse gases and fossil fuels. If anyone expressed any doubts about the safety of treating food with gamma radiation, scientific experts could assure the public that the procedure only produced effects similar to cooking the food.

The only problem with the scheme was that the public still is afraid of radiation. Winning over the public was definitely going to be a more serious matter than the DOE had reckoned. The department also had to face another sobering fact. America in the eighties was not the same place it had been in the fifties. Before, the DOE (or its predecessor, the Atomic Energy Commission) had had carte blanche when it came to anything atomic. The public didn't have any firsthand experience with radiation, Americans trusted their government, and there was a fervor about bomb-making and national security. With time and experience, public attitudes began to change. People no longer blindly trusted the .OE, or any government agency for that matter. Members of the public started to ask questions, and what was worse was that, when they got answers, they usually didn't believe them.

This public skepticism has been disappointing to the DOE, which still insists that food irradiation is a capital idea. On its own initiative, the Food and Drug Administration (FDA) reviewed the procedure, and with unprecedented efficiency, approved food irradiation as safe. In state after state, public opposition has defeated irradiation legislation. In May 1987, Maine became the first state to ban the sale and distribution of irradiated food. Two years later, New York governor Mario Cuomo signed into law similar legislation prohibiting irradiated food in that state.

Legislation allowing irradiation has been introduced and quietly passed in some states without the public even knowing the matter was up for discussion. While we believe that the legislation adopted by Maine and New York will eventually become models for other states, this will only occur as a result of citizen vigilance and awareness.

Is Food Irradiation Safe?

It is currently believed that food does not become radioactive when exposed to ionizing radiation, but there is considerable controversy about the safety of the resultant product. Proponents of irradiation claim that the procedure is similar to cooking food. The Task Group for the Review of Toxicology Data on Irradiated Foods, commissioned by the FDA to study the safety of irradiated foods, also concluded that studies appeared to support safety and that food irradiated at doses not exceeding 100k rads is wholesome and safe for human consumption and should be exempt from any toxicology–testing requirements. From these conclusions, the Food Marketing Institute agreed that most toxicologists and pathologists consider irradiated food to be as safe and nutritious as nonirradiated food.

According to scientists at Food and Water, Inc., a nonprofit organization, when you look at what the FDA studies actually say, there is little basis for such beliefs. Concerning the report by the Task Group for the Review of Toxicology Data on Irradiated Foods, conclusions were founded on a review of animal feeding studies with irradiated food. Starting with 441 studies, the Task Group found that only 69 were complete enough to be evaluated. Of these sixty-nine studies, 32 indicated adverse effects and 37 appeared to support safety. But on detailed examination of these 69 studies, the task group found only 5 studies appeared to support safety, and all the remaining 64 were determined to be deficient. They also noted that major questions still exist as to the safety of irradiated foods treated at high irradiation doses that constitute major contributions to the daily diet for long-term use.

Irradiation proponents have long claimed that the chemical changes taking place within food that is irradiated are similar to those of cooking. But, the State of New Jersey Report on Food Irradiation says, "This is clearly not the case.... Obviously, if irradiation and cooking created similar chemical changes, there would be no need to irradiate; the food could simply be cooked. A cooked steak is hot, browned, tender, and has aromatic qualities. An irradiated steak is cold, raw, indigestible and has a different aromatic quality. A baked potato left at room temperature would begin to decay in two days. An irradiated potato will not begin to decay for two years. The reason? The chemical changes are completely different."

In addition to these obvious differences between cooking and irradiation, chemicals called radiolytic products (RPs) and unique radiolytic products (URPs) are created within the food itself when it is exposed to ionizing radiation. These are new chemicals created by radiation and exist only in irradiated food.

Irradiation uses ionized energy. This means that the molecules of the food are broken apart. These broken pieces are known as free radicals. When they regroup, they form new molecules (UR's). In microwave cooking (nonionized radiation), the molecules simply vibrate. They are not broken apart. They do not form new chemicals. Normally, in cooking with water, food is broken down into its natural components, i.e., proteins, vitamins, etc. This is similar to the human body's process of digestion. Cooked food can be considered predigested. No new chemicals are formed. Frying can create the carcinogen benzopyrene. In irradiated foods, the new chemicals are created throughout the food. The New Jersey report concludes, "Irradiation proponents argue that since carcinogens may be created by frying, we should allow them to add carcinogens to our food supply through irradiation. This would be analogous to advising a two-pack-a-day smoker to become a four-pack-a-day smoker as his original consumption was cancer-causing."

The FDA has adopted the position that the amount of these new chemicals in irradiated food is too small to be harmful. This position is based on the conclusions of the Task Group, which in turn relies on the July 1980 report of the Bureau of Foods Irradiated Foods Committee (BFIFC). According to that report, if food is irradiated at levels of 100k rad or smaller, RPs will be produced at 30 parts per million and URPs at 3 parts per million. These calculations are, says the Task Group, "based, in part, on considerations of radiation chemistry and physics." Consequently, they are extrapolated from theoretical assumptions and not based on actual experiments with irradiated food. It is on the basis of these findings that the Task Group says, "We therefore agree with the Committee's conclusion that food irradiated at doses not exceeding 100k rad is wholesome and safe for human consumption."

Even if these rather tenuous calculations are accepted at face value, there is little reason to assume that irradiated food is safe. According to the New Jersey report, top government scientists have stated that neither the RPs or URPs have ever undergone toxicological testing. It should be noted that the pesticide EDB causes cancer at the parts per *billion* level and the herbicide 245T causes cancer at the parts per *trillion* level. The assumption that 3 parts per million of an untested chemical is safe seems unfounded.

In congressional testimony on food irradiation, Dr. Richard Piccioni, Senior Staff Scientists for Accord Research and Educational Associates explained some of the problems with testing irradiated food for safety: "Treatment of food with ionizing radiation presents issues of food safety qualitatively unlike those posed by any other food–processing method or food additive. The large amount of energy contained in ionizing radiation provides the potential for exceedingly complex chemical transformations of food components, including the production of mutagenic or carcinogenic substances which were not present, or were present in far smaller amounts, before irradiation. This potential far exceeds that of ordinary heat processing, microwave radiation, etc., because the energy contained in each 'quantum' of gamma radiation is so great. It should be clearly understood that without toxicological testing at exaggerated doses, the carcinogenic risk to large human populations ingesting any additive or residue is impossible to assess."

Dr. Piccioni is also concerned about the way in which the FDA has gone about approving food irradiation. In particular, he points to "an extraordinary leap of faith" in the FDA's acceptance of assumptions that may have little basis in reality. The BFIFC acknowledged that feeding whole, irradiated foods to laboratory animals, even over long periods of time, cannot provide a meaningful assessment of the carcinogenic potential of the radiolytic products present in those foods. As an alternative to direct biological testing, they proposed acceptance of a theoretical calculation of the maximum concentration of radiolytic products present in irradiated food.

Proponents of food irradiation have made a number of claims about the process as though they were statements of fact founded on valid scientific ground. We have seen that the claims of safety, although purportedly derived from legitimate scientific investigation, are in fact conjecture (mixed, perhaps, with some wishful thinking). As noted by Piccioni, it is virtually impossible to test the safety of irradiated foods according to accepted scientific standards.

Moreover, according Dr. Donald Louria, chairman of the Department of Preventive Medicine and Community Health at New Jersey Medical School, the studies relied upon by the FDA for endorsing the safety of irradiation are not only seriously flawed, but in fact may actually indicate that irradiation is a health risk.

"It would appear, " says Louria, "that the FDA gave its approval on the basis of five or six studies on rats and dogs. These were selected as methodologically sound from a pool of over two thousand studies . . . Clearly there are many potential biases in choosing such a small number of studies on which to base major decisions." In one of the studies, published in 1964, there were differences between controls and those rats given irradiated wheat, but the small numbers of animals may not have permitted statistically significant differences to be found. There were unexplained stillbirths in the litters of rats given wheat irradiated with 20,000 rads; recalculation of that stillbirth rate shows a significant increase. This study is hardly an endorsement for the safety of irradiating foods. In another study, there are similar problems with statistical significance, unexplained deaths, and abnormalities in animals given irradiated foods that are treated dismissively and virtually ignored.

Instead of documenting safety, the studies raise questions about the safety of food irradiation. Additionally, one of the studies suggests that older animals may be more susceptible to adverse effects when eating irradiated foods.

According to Louria, these studies suggest another problem associated with irradiated food—namely alterations in the nutritional value of the treated food. In the 1964 study, says Louria, the animals fed with irradiated wheat were given nutritional supplements in part to avoid the reproductive difficulties that were attributed to the destruction of vitamin E induced by radiation. In a German experiment, animals fed on irradiated food showed reproductive problems and lower body weights than the controls. Both of these problems were corrected by giving vitamin supplements, including vitamin E.

This nutritional alteration leads us back to the claims of the proponents of food irradiation that the process is similar to cooking a food. This ignores the fact that many of the foods proposed to be irradiated would never be cooked. These include papaya, mango, and other fresh fruits and vegetables. Consumers would be under the false impression that they were receiving full nutritional value of foods that appear to be raw but are in fact, irradiated and had ies on the effects of irradiation on nutrient quality, the USDA found that when irradiated pork was cooked, the "degradation of thiamine was significantly increased... The two processes (irradiation and cooking) produced degradation; but when the product was cooked after it had been irradiated the overall effect was greater than the sum of the processes applied individually."

Also regarding the safety of food irradiation, proponents say that it has been studied for more than 30 years and that there are no studies indicating that it causes any harm. Many people argue with this assertion as well. According to the findings of the New Jersey report, FDA and army officials even admit that the history of food irradiation research has been filled with "fraud, incompetence and poor research techniques."

The army began researching irradiation in 1953. In 1963, after reviewing a *summary* of studies on irradiated bacon, the FDA approved the irradiation of canned bacon for the army. Upon review of the actual studies in 1968, however, the FDA rescinded their approval because, the agency concluded, data showed significant adverse effects produced in animals fed irradiated food." In the 1970s, the army contracted with an industrial testing laboratory to research irradiation of food. The studies presented a large body of research supporting the safety of irradiated food. This same company, however, later went to trial on charges of conducting fraudulent research. Four officials from the large research company were under investigation for allegedly misrepresenting test data and supplying false lab reports. The company's studies were considered invalid. The lab was also on contract to the EPA to test the safety of many pesticides. Although an EPA spokesperson conceded at the time that the records of the research company were shocking, the lab did nothing about the chemicals, many of which are still in wide use. The FDA showed a similar lack of concern. It reviewed all of the studies on food irradiation, found 99 percent of them to be invalid, but instead of banning irradiation until legitimate studies could be performed, the FDA announced that it would no longer require toxicological studies to support the safety of food irradiation. Its rationale was based on the theoretical assumptions of the BFIFC that only minimal changes took place in food when it was irradiated.

According to Dr. Piccioni, it is also untrue that there are no studies in the scientific literature showing mutagenic or carcinogenic activity in irradiated food. "In fact," he says, "dozens of such studies exist, carried out in a variety of biological systems, and published by a variety of authors in a variety of peer-reviewed scientific journals over a period of twenty years." Piccioni backs up his assertion by a list of more than 30 studies reported between 1961 and 1981 in which scientists observed many different types of damage resulting from irradiation—ranging from chromosomal damage to plant embryos when plant leaves are irradiated, to mutagenicity, chromosomal damage, lowered sperm count, autoimmune disease, and polyploidy in laboratory animals fed irradiated food. One study was conducted in India on malnourished children who were fed freshly irradiated food. The children developed abnormal blood cells called polyploids, which contain more than the normal set of chromosomes and are suspected of being linked with leukemia. Other studies on rats, mice, monkeys, and hamsters showed similar increases in polyploid cells.

Contrary to what Americans are being told, the safety of food irradiation is far from being scientifically proven . In fact, many people believe that scientific literature actually indicates that irradiation is dangerous. It is only through unprecedented manipulations of its own rules and regulations that the FDA has been able to declare food irradiation to be safe and wholesome. Why has the FDA gone to such lengths to legitimize a procedure of questionable safety and opposed by a large segment of the population?

What's So Great About Irradiated Food?

Because of people's innate distrust of radiation, one of the major promotional devices used to present the concept of food irradiation to America has been to prove its safety. But there has to be some reason to irradiate our food supply in the first place. Safety, in the case of food irradiation, is not a reason to adopt the process, but rather is a rebuttal of charges that it is dangerous. So what are these reasons? Why should we welcome the idea of irradiating the world's food supply?

One of the primary reasons for food irradiation, its proponents claim, is that it can extend the shelf life of food almost indefinitely by destroying bacteria that cause the product to spoil. From the point of view of the military, this can be seen as a desirable objective, especially in times of war. Also understandable is the approval of food processors and grocers—an extended shelf life means less spoilage, less waste, and more profits. But is this something that Americans really want and need for their general well-being? The growing trend in America is toward more fresh fruits, vegetables, and generally healthier foods. The large number of products on supermarket shelves labeled *No Preservatives* indicates that the public does not share the food industry's enthusiasm for extending the shelf life of its foods. While food irradiation will give shoppers the impression of getting fresh and wholesome foods, in reality many of these "fresh" foods may be quite old and stale.

In terms of the nutritional value of irradiated foods, the process of irradiation itself may destroy essential nutrients like vitamins E, C, and K, the amino acid methionine, while causing chemical changes in proteins, fats, and carbohydrates, the effects of which remain unknown. Vitamins and nutrients also deteriorate over time. The older the product, the less nutritious it becomes. Irradiation, while killing organisms responsible for spoilage, will not stop this natural deterioration of the nutritional value of food. Consequently, in the future where foods are largely irradiated, we may find that many of our traditional sources of nutrients—oranges for vitamin C, grains for vitamin E, carrots for vitamin A—contain little more than traces of the substances required to maintain health. While synthetic vitamins can always be added or taken as supplements, they rarely can replace the natural compounds that occur in our food in just the right combinations and dosages to ensure proper absorption and utilization.

Some critics also worry that irradiation may create a situation similar to problems experienced with pesticides where less tolerant beneficial insects (bacteria, in the case of irradiation) are killed off, leaving the more resistant pest/bacteria to thrive unchecked by its natural predators. Another concern is that the more harmful bacteria will mutate and become resistant to radiation, just as pests have developed resistances to pesticides. For instance, salmonella causes food to smell and taste bad when it spoils. The common bacteria *E. coli* is also responsible for signs usually associated with spoiled food. Both of these bacteria are killed at lower doses of radiation than the bacteria *C. botuliinum*, which causes botulism and requires very high doses of radiation to kill. This has a potential to create a hazardous situation; if the salmonella, *E. coli*, and other bacteria that cause spoilage are killed but the botulism bacteria are not, food can look and smell fresh but actually contain a dangerous amount of botulism toxin.

Dr. Piccioni points out that in order to inactivate salmonella in poultry, one of the prime irradiation targets, two points must be understood:

•Doses of radiation required to kill even the salmonella are larger that the safe dosages established by the FDA The levels of mutagens contained in foods irradiated at the safe level of 100k rads are only exacerbated at doses of 1 million rads, required for even partial salmonella inactivation.

•Much of the illness caused by salmonella poisoning has been attributed to resistant strains of the bacteria developing in response to the wide use of antibiotics in the poultry industry. "The addition," says Piccioni, "of a highly mutagenic processing procedure, namely, gamma irradiation, on poultry carcasses still containing low levels of antibiotics is an alarming scenario for the appearance in the irradiated food of new, antibiotic-resistant strains."

Because irradiation retards spoilage, often for many years, another reason given to support its use is that it will enable us to feed the world's hungry. No one would argue the importance of attaining this end, but there is little reason to believe that food irradiation, or any food–processing scheme, is going to take us any closer to solving this problem. Many times politics plays a much more important role in thwarting efforts to feed the hungry than a lack of food. We now hold silos full of surplus grains and pay our farmers to stop producing foods. The problem of hunger both in America and abroad is not one of inadequate supply. Food surpluses rarely result in the hungry having more to eat. In Ethiopia, during the famine that gained so much world support, the problem was not a food shortage. Food was flooding into the country from around the world. As in many developing countries with food shortages, the Ethiopian hungry continued to starve because of political corruption and inadequate or nonexistent infrastructure (roads, distribution agencies, etc.). To claim that food irradiation is going to feed the world necessarily implies that it is capable of solving all of the political and sociological factors contributing to world hunger.

Another interesting argument in favor of food irradiation is that it will replace pesticides. This is a particularly attractive lure to people who are bothered by chemical pollution of our environment and our food supply. Food irradiation, however, is a postharvest treatment, and, consequently, will have no impact whatsoever on preharvest pesticides used, for example, to fumigate and sterilize soil, as fertilizers or as insecticides applied while the plant is growing. Arguably, irradiation could be used to replace some postharvest fumigants, but even then these chemicals would still be used liberally until the food actually makes it to the irradiator. What this all means is that instead of replacing pesticides, in most cases food irradiation will add yet another step to the treatment of our food. "In fact," says Piccioni, "irradiation of fruits and vegetables may well increase, rather than decrease, the requirement for postharvest application of fungicides because irradiated products are more susceptible to infection by molds and fungi."

An issue at the fore with the critics of food irradiation is the environmental and transportation hazards inherent in a large-scale irradiation program. The program proposed by irradiation supporters would entail hundreds, if not thousands, of facilities across the country, often in the heart of high-population urban areas. The safety records of both military nuclear facilities and civilian–power reactors have shown us that nuclear technology presents grave dangers to both the environment and human health. The irradiation facilities proposed by the DOE would utilize approximately 3 million curies of radioactive material. To gain some perspective on the environmental significance of the quantity of isotopes in a large irradiator, consider the fact that the Chernobyl reactor accident, which continues to cause significant agricultural disruption many hundreds of miles from the reactor site, resulted in measurable levels of CS-137 in air in New York City, involved the release of 1 million curies of CS-137, one third the planned inventory of DOE irradiation facilities.

In its report on food irradiation, Food and Water, Inc., notes that "food irradiation technology presents major environmental considerations similar to those posed by other nuclear processes but in some respects more severe because of the large number of activities involved, the high-level radioactive sources at each facility, and the lesser degree of regulatory control required." The report outlines some of the major concerns:

1. Handling and transportation of large quantities of radioactive materials for a widespread commercial activity like food irradiation presents serious possibilities of contamination in the event of an accident;

2. Increased exposure of workers to high levels of radiation. The report cites the 1977 accident at Radiation Technology, Inc., where a worker opened a door to a radiation chamber while the irradiation process was in effect. The worker received a large but, in this case, not lethal dosage of radiation;

3. Contamination of the environment and groundwater due to lax safety precautions. This problem is all too evident in areas surrounding the DOE's weapons–making installations, where the department knowingly released radioactive materials into the environment for years. Groundwater around many of these sites is now contaminated. The report cites two known instances of this occurring at irradiation facilities. Once in 1976 at the Isomedix plant in Parsippany, New Jersey, and another time in 1982 at the International Nutronics plant in Dover, New Jersey, where radioactive waste water was poured into the sewer system;

4. The release of radiation-resistant mutant bacteria into the environment;

5. Adding to the problems and costs of nuclear–waste disposal. Additionally, says the report, "Not the least among the environmental hazards...is that with the creation of new companies as this multi-billion dollar 'growth industry' get under way, the likelihood is markedly increased" that unscrupulous entrepreneurs will illegally and unsafely dispose of their radioactive wastes. Midnight dumping of hazardous wastes is well documented;

6. With widespread usage and transportation of radioactive materials, the chances of terrorist threats increases. A possible scenario cited by the report is a time bomb being placed in a food crate. Exploding in an irradiation chamber, large amounts of radioactive materials could be released into the environment. Common circulation of radioactive materials also increases access to materials required to make nuclear bombs.

7. Accidental explosions could release large amounts of radioactive materials. The report points out that many of the irradiation plants under consideration are large centralized facilities in the heart of metropolitan areas. Evacuation of large populations in the event of an accident would be virtually impossible.

These concerns are not the fabrication of an alarmist fringe. The midnight dumping of hazardous materials mentioned above is now recognized as a major cause in some of the country's worst toxic–contamination sites. It is costly to deal with waste, particularly radioactive waste, in a responsible and effective manner. Already we have instances where corner-cutting, corporate fraud, and lax safety regulations exist within the irradiation industry. A notable case is that of Radiation Technology Inc., the company cited above for the worker–safety accident. In March 1988, the New Jersey–based company, which operates irradiation facilities to sterilize spices and medical equipment, pleaded guilty to charges that it had falsified reports to the Nuclear Regulatory Commission concerning safety procedures at its plant. Guilty pleas were also entered by top company officials involved in the conspiracy.

History and Politics

Just about everything having to do with nuclear technology in this country has a strange and often hidden side to it. Historically, this is explained by the secrecy surrounding the development of the atomic bomb, followed by the Cold War. From the outset, the atom forged a strong bond between the military and those sectors of government and society having to do with the production of energy—hence the nuclear weapons making plants are under the control of the DOE (and not the Defense Department as one might logically assume), while many of these facilities have been managed by companies like General Electric and Westinghouse. These two electrical giants, in turn, have major contracts with the DOE and the Pentagon for the manufacture of various military goods.

This intermingling between the military complex and the commercial sector, while to the obvious advantage of both, was, at the same time, part of a larger scheme—the Atoms for Peace program initiated under the Eisenhower administration. Early on, those involved with nuclear technology realized that public apprehension about the atom could unduly interfere with what they perceived to be the appropriate development of the new technology. The primary objective of Atoms for Peace was to find ways to make the public feel more comfortable about nuclear power. Since the fifties, a series of legislation has been enacted concerning the use of atomic energy. Their names alone indicate that, besides use and development, waste disposal and associated costs were of growing concern to both the DOE and the Congress—the Energy Reorganization Act of 1974; the Nuclear Waste Policy Act of 1982; the Low-Level Radioactive Waste Policy Act of 1980 and its 1985 amendments; and the 1987 Budget Reconciliation Act. In addition to its responsibility for the production of the nation's nuclear–weapons arsenal, Congress also explicitly appointed the DOE as the federal agency responsible for the disposal of all high-level nuclear waste (i.e., that generated in both nuclear weapons–plants and commercial facilities). As part of its responsibilities in this regard, in 1985, the DOE initiated the By-products Utilization Program, now called the Advanced Radiation Technology Program, to research and develop different ways of using radioactive waste.

"Among the by-products, or wastes that DOE seeks to utilize," says Judith Johnsrud of Food and Water, Inc., "are strontium 90 (which concentrates in bone and has been long recognized as a biologically hazardous isotope from atmospheric nuclear weapons testing in the fifties), tritium, xenon, plutonium-238, and most significantly, cesium-137 for food irradiation. Although individuals in the equipment– irradiation business insist that their plants will use only cobalt-60 for food irradiation, Congress appropriates $5 million annually to DOE to demonstrate that cesium-137 will do the job, and at a lower cost than imported cobalt-60."

It is difficult to discern whether the DOE is acting independently to promote the commercial utilization of radioactive wastes, or whether projects like food irradiation are part of general government policy, as enunciated by the 1954 Atomic Energy Act. Since Congress is appropriating funds for research, it must be at least in tacit approval of the project. A high-level policy decision would also explain why the FDA acted as it did in approving food irradiation.

Critics of food irradiation suspect that the FDA's unusual behavior, as well as many of the other peculiarities surrounding the irradiation program, may be explained if seen as part of a concerted government effort led by the DOE—the main proponent of food irradiation.

In discussions about what radioactive sources are to be tapped to fuel a large-scale food–irradiation program, proponents typically propose using Canadian cobalt–60. But, as pointed out by the New Jersey report on food irradiation, it is naive of the irradiation industry to suggest that the DOE is spending millions of dollars to promote irradiation so that the industry will buy cobalt 60 from Canada. Furthermore, Piccioni believes cobalt 60 production will be overwhelmed by the demands of a large food irradiation industry in the United States and elsewhere. The only practical alternative is cesium-137, a nuclear–waste material.

According to a letter from F. C. Gilbert of the DOE to the House Procurement and Military Systems Subcommittee in March of 1983: "The strategy being pursued by DOE's By-products Utilization Program is designed to transfer federally developed cesium-137 irradiation technology to the commercial sector as rapidly and successfully as possible. The measure of success will be the degree to which this technology is implemented industrially and the subsequent demand created for CS-137."

Generating a large demand for cesium-137 solves a number of problems for the DOE. First, cesium-137 is one of nuclear waste's most radioactive components. According to DOE sources, cesium-137 comprises more than 50 percent of the radioactivity and of the heat generation, but only 3 percent of the volume of nuclear waste at the Hanford, Washington, nuclear facility. The DOE has estimated that it could save $1 billion on storage and disposal of radioactive waste if the cesium, strontium, and americium were first removed. According to Johnsrud, in the early 1970s, DOE's predecessor had simplified the storage of high-level wastes at the Hanford site by removing cesium and strontium from defense wastes. Those purified and encapsulated cesium wastes are now to be utilized, under the DOE By-products Utilization Program to demonstrate the efficacy of food irradiation.

The DOE will save money, but will also turn a profit when it sells its cesium to commercial food irradiation. Cesium and cobalt were priced at $1 per curie. However, the DOE has lowered the price to 10 cents per curie to make their waste attractive. Both materials do the job equally well. It is necessary to use 5 times more cesium, which brings the cost to 50 cents per curie—still better than a $1 per curie for cobalt. Also cobalt has a half life of 5 years; cesium has a half-life of 30 years. This means that using cobalt would require investing more money every 5 years to replenish the source, as opposed to every 30 years for cesium.

Under federal law, the DOE is responsible for the disposal of all high–level radioactive waste—generated both from its nuclear–weapons plants and also from commercial uses. In 1980, Congress determined that states would have to bear the responsibility for low level wastes originating in each state. What concerns Johnsrud, and many other observers of the food–irradiation debate is that while manipulating the definition of what constitutes high–level versus low–level radioactive waste may be politically expedient, it hardly provides an adequate response to the waste–disposal problem. Cesium still remains a highly radioactive material and requires carefully designed disposal procedures. With the initiation of a largescale irradiation program cesium will become virtually ubiquitous. If states and private companies are given the discretion to dispose of the troublesome waste by simply diluting it to conform with minimum standards and then dumping it down the drain or into the municipal dump, this could lead to a nationwide contamination of the environment.

This leads to the next piece of the puzzle. Since there is not enough cobalt and not even enough cesium available at government weapons facilities to fuel a large-scale irradiation program, it seems logical to start the maverick project out on smaller scale to see how it works. That would give scientists the time to evaluate its efficacy at doing what proponents claim it is capable of doing. Health and environmental effects could also be studied before billions of dollars were spent on a project that may have to be abandoned a few years later when experience shows that its risks outweigh its benefits.

Keeping Your Food Nuke-Free

Food irradiation is admittedly a very confusing issue. Common sense certainly does not help in coming up with a reasonable explanation as to why it is being backed so strongly. If we told you, for instance, that we had this brilliant new idea and explained food irradiation to you for the first time, chances are you would think we were quite mad. Your thoughts on the subject might be something like this: "Why would you want to expose our food supply to *radiation*? That stuff is toxic, it causes cancer and mutation. Expose our food to it? What are you talking about?"

We believe that this is the common sense reaction to food irradiation. Even if it could do all the things its proponents claim it capable of—extending shelf life, feeding the world, eradicate the need for pesticides—food irradiation is still a fundamentally dangerous and undesirable process. What Americans want and need today is fresh, wholesome, and nutritious food. The food producers and suppliers of this country need to recognize these consumer demands and take a couple giant steps *backwards* in time rather than offering us atomic food. All of our medical research is showing that we should be eating more fresh, whole–grain, and unprocessed foods, less fat, and less meat. This is the type of diet that people have been eating for hundreds of years. Apart from food shortages and dietary imbalances due to poverty, there was very little dietary-related illness in societies living on these foods. The epidemics of cancer, heart disease, atherosclerosis, and arthritis are a modern, primarily American, phenomenon. And many of them are directly related to our processed foods that are not only fiberless, fatty, and sugary, but also devoid of the essential vitamins, minerals, and enzymes required to maintain health.

What we need now for our health and well-being is food that is *really* fresher and more nutritious. Food irradiation is a mask that serves to confuse and mislead the consumer. It can make old foods appear fresh and healthy, when in fact they are years old and devoid of any nutritional value. Consumers who buy fresh produce do so because they are conscious of their health and want to buy the best sources of food to promote it. Irradiation of fresh fruits and vegetables would make these attempts to maintain health a sham.

Undoubtedly the proponents of irradiation realize that the new atomic food is not going to be an overnight success with consumers, for they have vehemently opposed any labeling requirements. Initially, the FDA proposed that there be no labeling on irradiated food. Consumer opposition caused them to change their minds, but, says the New Jersey report, "The current FDA rule for irradiated food is puzzling." While the FDA requires labeling for whole irradiated foods, no label is required on processed foods using irradiated ingredients. So, irradiated flour would have to be labeled, but bread made from that flour would not.

Moreover, even these flimsy labeling requirements were scheduled to expire on April 18, 1988, after which irradiated food would be marked only with a symbol resembling a tulip in a circle. Again, the New Jersey report indicates the deception involved: "In 1986, prior to the FDA approval, both the irradiation industry and the National Food Processors Association estimated that the marketing and production of irradiated food would take at least two years. In other words, the FDA knowingly issued a rule requiring the labeling of irradiated food for a period of two years when it was fully aware that no irradiated food would be on the market. By the time irradiated food is introduced to the market, the Federal labeling requirement will have expired...In reality, the labeling requirement is meaningless since the FDA has practically no labeling enforcement program."

Under public pressure, the FDA did rule to extend the "Treated with radiation" or "Treated by irradiation" labeling requirement for an additional two years. After that, irradiated food need only be marked by the irradiation logo, which, despite protests, is markedly similar to the identifying logo of the EPA. Look for this information on all labels and wrappers when you shop. Ask your grocer if his produce has been irradiated, and shop elsewhere if he doesn't give you a straight answer. Organize other consumers in your community to boycott stores that carry irradiated foods. Some communities are getting the idea: New York recently passed a law that bans all irradiated foods (with the exception of dried spices and foods prepared for hospital patients with immune deficiencies) from sale within the state. Unfortunately, Maine is the only other state to have such a law in effect.

Any way you look at it, there seems to be very little for consumers in this irradiation business. They get old, stale food that contains unknown quantities of substances with suspected carcinogenic and mutagenic effects. The nation is already floundering in its attempts to deal with 40 years of atomic technology and waste. A large-scale irradiation industry would not solve, but compound these problems. Nuclear contamination would no longer be confined to areas surrounding weapons facilities and power plants, but could spread into the hearts of cities, down rivers, into oceans. We have seen the problems with oil spills over the last few years. These may seem like child's play compared to nuclear accidents resulting from things like ambulatory irradiation facilities moving from field to field, and across the nation's roads, to conduct onsite treatment of foods.

When you look at who is supporting food irradiation, the issue takes on an even more confusing aspect. Why should the DOE have anything to say about what America eats? What is the FDA, whose primary mandate is to ensure the safety of our food supply, thinking about when it ramroded through its approval of irradiation? The Reagan administration, strong on exotic weapons programs like Star Wars, was also a supporter of food irradiation.

Is this all mere coincidence? Are the benefits to be reaped by the proponents of irradiation just unexpected windfalls of a project originally designed to promote public health and well-being? Despite the importance of safety issues—both with respect to human health and the environment—many feel that the plutonium connection is by far the most dangerous aspect of irradiation.

Part Three
The Waste We Leave Behind

Chapter 9
Nuclear Waste

On April 28, 1986, the most feared catastrophe in the 32-year history of commercial atomic power took place when a chemical explosion destroyed a nuclear reactor at the Chernobyl nuclear power plant, 80 miles north of Kiev, the Soviet Union's third–largest city. Although the reactor was of Soviet design and the core did not actually melt, in terms of radioactive fallout, the Chernobyl disaster was easily equal to the "worst-case" accident at a Western-style water-moderated reactor—a "meltdown." While the short-lived radioactive substances had already decayed at the time of the accident, long-lived radionuclides such as cesium-137 (half-life of 30 years), strontium-90 (27.7 year half-life), and plutonium–238 (half-life of 24,000 years) remained and were released during the accident. Soviet officials infuriated the world with a delayed and tight-lipped reaction, and left nuclear experts to surmise what took place without firsthand facts.

Scientists now believe that a loss of water coolant caused the temperature in the reactor's core to soar upwards of 5000 degrees Fahrenheit. Uranium fuel probably began melting, producing steam that reacted with the zirconium alloy coating of the fuel rods and producing a highly explosive hydrogen gas. Another reaction between steam and graphite produced hydrogen and carbon oxides that, combining with oxygen, blasted off the top of the building that housed the reactor, and ignited its graphite core. A dense cloud of radioactive–fission products was spewed into the air. Scientists now estimate that 100 percent of the inert gases, as much as 50 percent of the iodines, telluriums and cesiums and 3 to 6 percent of all other materials in the core were released during the explosion. Burning at temperatures twice that of molten steel, fire fueled by the white-hot graphite was uncontrollable for a week because of the blazing temperatures and the extreme radioactivity.

It took 36 hours to marshall 1,100 buses to evacuate the 25,000 residents of Pripyat, the town where the crippled reactor was situated. During that initial day and a half, the townspeople of Pripyat and surrounding areas were subjected to very high doses of radioactive material. During a typical Xray, a person receives approximately one-fiftieth of a rem of radiation; nearly 100,000 people belatedly evacuated within twenty miles of Chernobyl received up to 10,000 times that dosage. At a very close range, plant workers and townspeople received even higher doses that within a few hours of exposure caused nausea, vomiting, cerebral hemorrhage, and often death. Between one and three miles from the reactor, people suffered a massive suppression of white blood cells, blotched skin, and acute bowel syndrome, with an 80 to 100 percent chance of death within four to six weeks. Those living within three to four miles of the site would have a 50 percent chance of dying as their bone marrow, highly sensitive to radiation, was damaged, thus reducing the body's ability to fight infection. Those surviving would more likely than not suffer permanent damage to their bone marrow and gastrointestinal tract. People living five miles or more from the reactor would probably not die soon after the accident, but they would be subject to a greatly increased risk of contracting leukemia and other forms of cancer. Genetic abnormalities could be expected from damage to the reproductive organs.

But human health was not the only thing to be seriously damaged in the wake of the Chernobyl accident. Farmland for many miles around the city was covered with radioactive particles that could take decades to decay, while groundwater supplies for more than 6 million people could be irreparably contaminated if the white-hot graphite core burned through the underlying layer of concrete. The particulates and gases released from the plant blew a mile-high plume that was carried by prevailing winds to most of eastern Europe, inciting panic and anger as governments frantically proposed safety measures.

The accident at Chernobyl gives us some idea of what a "worst-case" scenario can look like. It was the largest release of radioactivity ever recorded in a single power–plant disaster. In terms of cesium-137, one of the most potent components of radioactive fallout, the accident at Chernobyl released radioactivity equaling several dozen Hiroshima bombs.

The European community has issued a report that estimates the additional cancer deaths caused by the Chernobyl to be no more than about 1,000 over the next 50 years and reassures people that their average exposure to radiation did not exceed that of normal background radiation. As we will discuss later on in this chapter, there are other scientists who believe that the legacy of Chernobyl has been grossly misunderstood and underestimated. In the United States alone, Drs. Ernest Sternglass and Jay Gould believe that the effects of radioactive fallout from Chernobyl have already caused an additional 40,000 to 60,000 American deaths. Nor are exposure levels confined to the minimal amounts assured by the government officials. In 1988, two years after the accident, a sheep farmer in Northern Wales was found to have 4,118 becquerels of radioactivity per kilogram of body weight—about 10 times the normal amount. His wife was found to have 9 times the normal amount. There are other reports that belie the official statements:

•In Britain, 300,000 sheep cannot be slaughtered because their bodies contain dangerous levels of radiation;

•Reindeer in Sweden must be transported in huge trucks for grazing in order to avoid contaminated pasturage;

•Fishing is still banned on Switzerland's half of Lake Lugano, and people are still wary of many fish.

Although it is the worst nuclear accident known to have taken place, the Chernobyl incident is not the first. Since the first nuclear power plant began operating in the Soviet Union in 1954, a series of nightmarish close calls have plagued the industry at an increasing rate.

Soviet officials maintain that the cause of the catastrophe at Chernobyl was human error. But how reassured should that make us feel since, at least for the time being, our reactors are operated by human beings? In fact, most of the serious reported nuclear accidents have been linked to human error—and not just in the Soviet Union. In 1952, four control rods were mistakenly removed from the Chalk River nuclear reactor near Ottawa, Canada. A partial meltdown caused the accumulation of a million gallons of radioactive water inside the plant. Another worker at a reactor near Idaho Falls erroneously removed control rods from that reactor's core in 1961. A resulting steam explosion killed three workers. In 1975, a fire was started at the Browns Ferry reactor near Decatur, Alabama, by a worker using a lighted candle to find air leaks. Electrical cables connected to the cooling system were damaged, and the reactor's cooling system was dangerously impaired. In the most serious of American close calls, human error was again responsible for a loss of coolant at the reactor at Three

Mile Island, near Harrisburg, Pennsylvania, in 1979. A partial meltdown occurred and radioactive material was released.

Nuclear Mythology

America entered the nuclear age in the midst of a frenzy of patriotism. The time was World War II. The cause was the Manhattan Project and the development of the Bomb. The purpose was to defend America from the Japanese and the world from Hitler. America's top scientists, corporations, and government officials all rallied in a joint effort to keep the world free. National security took top priority and secrecy reigned around the research and development of the bomb.

Even with the passing of more than 40 years since the development of the atomic bomb, many vestiges of its wartime origins continue to surround the issue of nuclear power and weaponry. In probing under the surface of the nuclear establishment, we can still witness an alliance between the federal government and some of the country's most influential corporations and institutions. There is yet a massive influx of government funds to keep nuclear policy alive and headed in the "right" direction, while the secrecy that continues to dominate in and around the nation's power and weapons facilities is still explained as being necessary for national security. For this reason, it is important to look at some of the claims proponents espouse and to try to discern how closely they align with reality.

Myth No. 1: Nuclear Power Is Inexpensive
Ever since nuclear fission was first introduced to the American public as a viable source of power for our energy-hungry nation, one of the major arguments in its favor was that it was inexpensive. In fact, during the initial campaigns in the late 1940s and early fifties, the chairman of the Atomic Energy Commission held out nuclear as an energy source that would be "too cheap to meter."

We now know that this is fundamentally untrue. This is not to say that those who wanted to introduce the nation to nuclear made their claims in bad faith. In the early days of the atomic age, very little was known about nuclear power. Its risks and its costs were just not fully understood. But today, after many years of experience, it is apparent that any promises that nuclear power can provide an inexpensive source of energy for the nation or the world are at best naive.

One of the primary reasons that proponents of nuclear power can continue to aver its low cost is that many important pricing factors are conveniently omitted. For instance, rarely taken into account when discussing the costs of nuclear energy is the cost of development which was almost totally subsidized by the federal government in its wartime rush to develop the Bomb.

The costs of research and development, even though they have been largely omitted from discussions by the nuclear industry in the United States, are nevertheless very real costs. They are costs that must be borne one way or another—either indirectly through taxation or directly through utility rates. And it is precisely this type of hidden cost that makes nuclear power seem feasible today when other alternatives to fossil fuels such as solar energy seem not to be. There is little doubt that if solar or other forms of renewable energy had received the government support both in terms of enthusiasm and financial support these forms of energy, not nuclear, would be the viable and inexpensive alternatives to fossil fuels that nuclear is touted to be.

Other questionable accounting methods have been designed to make nuclear seem more inexpensive than it really is. In Britain, for instance, up until recently the costs of nuclear-generated electricity were decreased by what was called a "plutonium credit." Since plutonium was produced as a by-product of nuclear–power generation and was also required for the manufacture of nuclear weapons, it could be sold for military use in bombs or for further civil use. When it became evident that no plutonium market existed, however, the credit was first reduced and then altogether phased out.

Two other practices commonly used in understating the real costs of nuclear power are the use of inflation to make capital–investment costs appear smaller than they really are and discounting long-term costs to also make them appear much lower.

Discounting is an accounting procedure that allows for a comparison between future costs and present benefits. Suppose, for instance, that I have a cleanup job to do in 10 years that will cost me $1,000 at that time. I may provide for that expenditure today by depositing a smaller amount in the bank or in a mutual fund that offers me a given rate of interest. While this may be a valid accounting method for short-term projects or in situations where I do indeed set the money aside and the costs are precisely as I estimated them, there are a number of flaws in using it for nuclear accounting. First of all, nuclear costs are very often spread over a long period of time, so that while one generation may reap the benefits, future generations will be left with the cleanup bills. Theoretically this would present no real problem if enough money had been set aside to cover the cleanup costs when they arise. But this has not been done. Furthermore, as we gain increasing knowledge about the problems associated with nuclear–waste disposal, there is ample reason to doubt that any amount of money can adequately compensate future generations for the environmental devastation that will threaten them because of it.

In January 1989, the Department of Energy estimated that it could cost between $53 billion and $92 billion to clean up the radioactive and chemical pollution at the bomb–manufacturing and civilian facilities under its control. These costs do not even include the estimated $1.8 billion annual expense of keeping the facilities in compliance with safety regulation, nor the expense of cleanup, maintenance, and decommissioning of nuclear power sites around the country not under the control of the DOE. Even so, the estimates drew immediate criticism. According to Senator John Glenn, who requested the study, "[I]t is clear that the Department had underestimated the costs." He added that the DOE had a history of downplaying the costs of its projects. While these figures relate primarily to the government's nuclear–weapon–making facilities, the costs involved apply equally to electrical generating plants as well. They too must be "decommissioned" when they are abandoned or worn out and their toxic waste must be transported and stored.

Needless to say, the DOE's cleanup estimates, which many Congress members believe could go as high as $200 billion, comes as quite a shock. We thought we had already paid for these weapons plants, and suddenly we find out that retiring and cleaning up after them may prove to be just as expensive as building them in the first place.

A case in point is the Special Metallurgical Building at the Mound Laboratory in Miamisburg, Ohio. A corrugated metal building about the size of a normal milking barn, the structure was built in 1960 to process plutonium. The construction took five months and cost $1 million. Now that the DOE no longer needs the facility, it is tearing it down. Decommissioning is scheduled to take 29 years, ending in 1997, at an estimated cost of $57 million. But the Miamisburg site is not alone. Thousands of contaminated sites that the DOE has accumulated since 1942 when the Manhattan Project was first undertaken have now been abandoned and await decommissioning. These problems are not confined to military installations. The DOE is in the midst of a $100 million project to decommission the first civilian power plant at Shippingport, Pennsylvania. Some other DOE projects include:

•A $17 million program to decommission 324 buildings and reactors at weapons facilities throughout the country. The department estimates that it will take until 2012 and cost $817 million to complete the first 45 projects;

•A $46 million-a-year project to decommission another 32 sites. The cost is estimated at $1 billion and is expected to take 20 years;

•A $22 million-a-year project to decontaminate laboratories and other facilities where nuclear research was conducted in the 1940s and fiftiess. Estimated cost is at least $960 million;

•A program established by Congress in 1978 to clean up and stabilize huge deposits of radioactive waste, also called mill tailings. These tailings also found their way into building materials such as concrete and mortar for private homes. The DOE spent $116 million in 1989 alone on this project. The department says that it can complete it by 1994 for about $993 million.

The magnitude of the job involved simply to clean up the nation's weapons factories prompts many in Congress to fear that costs may be so high that cleanup efforts may never get under way. Engineers at the Energy Department have privately begun calling such contaminated sites "national sacrifice zones." They also say that failing to address the issue could mean that contamination continues to spread through the environment.

Additionally, although the navy is being particularly secretive about its costs of decommissioning nuclear submarines, some engineers estimate that the cost per vessel could run into the millions of dollars.

This discussion of the costs involved with nuclear technology, particularly those relating the disposal of radioactive waste and decommissioning of nuclear facilities, ties directly into the second myth about nuclear technology.

Myth No. 2: Nuclear Is Safe
As we have already seen from the accident at Chernobyl, nuclear power is anything but safe. Nor is the laying of official blame on human error of much consolation, for it turns out that human error is the major source of mishaps and accidents at nuclear facilities. According to a recent study conducted by the Critical Mass Energy Project of the Public Citizen Research Group, an organization af-

filiated with Ralph Nader, errors by nuclear–plant personnel played some role in 74 percent of all mishaps at civilian facilities. The group cited the lax performance of the Nuclear Regulatory Commission, which monitors civilian reactors, in failing to issue regulations on safety and training programs as required under a 1982 law, enacted after the Three Mile Island accident. In explaining the importance of safety training at nuclear facilities, Kenneth Boley, author of the study, said, "If things are going fine, but then a pipe bursts, and an operator pushes the wrong button, that mistake keeps the accident going." This is precisely what happened at Three Mile Island when a relatively minor mechanical failure led to a near core meltdown due to operator mistakes.

Safety problems from human error are only part of the problem. One after another we are seeing our nuclear–power plants and bomb manufacturing facilities closing for safety reasons. Many parts of the bomb–production complex date to the 1940s and are built with little emphasis on maintenance and safety. These aging monoliths are closing one by one as their threats to human health and the environment are brought to the public's attention.

• 1987—The Hanford Reservation facility, the nation's biggest plutonium producer and the facility at which the plutonium for the Nagasaki bomb was made, was closed for safety reasons as well.

• April 1988—At the Feed Materials Production Center in Fernald, Ohio, a report released indicated "pervasive" management problems and worker exposure to radiation. The report also revealed that for many years radioactive waste had been stored in leaking containers, contaminating groundwater in the area, and that about once a month filters on the smoke stacks broke, allowing clouds of radioactive uranium dust to be released.

• August 1988—The Savannah River Plant in South Carolina, which has three reactors for the production of tritium, a vital and perishable component of nuclear weapons, saw all three reactors close down for safety reasons.

• October 1988—The plutonium–processing site at Rocky Flats, Colorado, was closed by an emergency order issued by the DOE because of an accident involving radioactive contamination of employees following a plutonium spill.

In October 1988, the government reversed its long–standing policy of denying any risk to people from nuclear facilities and admitted that even small amounts of radioactivity could be hazardous to the health of the people living close to the Fernald, Ohio, plant. While this admission represented a step in the right direction, the people of Fernald were understandably upset. According to documents filed in court the previous month, the DOE admitted that it had known for decades that "normal operation fo the Fernald plant would result in emission of uranium and other substances," but it did not want to spend the money on pollution controls. Said Doris Clawson, a resident whose family had lived for five generations on a large farm bordering on the plant, "The government was lying to us, and they lied and lied." Her husband, Marvin, expressed similar disillusionment with the way in which the government had treated them. "Those are the worst people you can imagine," he said. "You'd think we had some kind of enemy down there trying to do away with us."

Not only did the federal government withhold information that could perhaps have helped people to protect themselves if they had received adequate warnings, the official position had also been that there existed a "threshold" level of radiation below which exposure was harmless. Lisa Crawford, a spokesperson for the Fernald Residents for Environmental Safety and Health, said, "They've always told us that it's such a low level, that it won't hurt you, so it's shocking that they're now telling us a different story." The only thing that is unusual about the Fernald plant is that the government made any admission of risk to human health at all. In the meantime, the Westinghouse Electric Corporation, which has operated the plant since 1986, insisted that human health was not endangered by the plant. This statement was made notwithstanding reports from Ohio's Environmental Protection Agency that about 12.7 million pounds of uranium waste had been generated by the plant since its opening in 1951 and that an additional 167,000 pounds had been discharged into the Great Miami River and another 289,000 had been released into the air. Observing the Fernald situation in retrospect provides some important insights into the type of rhetoric in which government and industry engage when they have their own objectives to achieve. For instance, Michael W. Boback, the plant's health and safety manager, wrote in a letter to the residents of Fernald around 1983 that "the amount of airborne uranium which goes beyond" the confines of the Feed Materials plant "will, in time settle out on the ground and add to the natural soil content." While statements such as these may seem preposterous by today's standards, it is important to keep them in mind when assessing the validity of statements made today as to the safety of nuclear energy.

Just as the initial campaigns to launch nuclear power claimed it was inexpensive, those early promotions touted its safety. In order

to get people to overcome their fears about the testing and manufacturing of the bombs, a picture was presented of a technology so
safe that no one would mind having a power plant in their backyard. While the American public has become too savvy to accept
these ludicrous claims of safety, the government tenaciously ad-
heres to its stance that no one is ever harmed. And for the most
part, the press parrots these assertions. In December 1988, the DOE
released a report that found that numerous nuclear weapons sites
throughout the country posed serious threats to human health, the
most threatening situation being at the Rocky Flats site near Denver at which toxic chemicals had contaminated an underground
water reservoir and endangered the entire city's water supply. But
press reports claimed "No harm to people is found" and "Health
surveys have never shown any harm to people from the by-products of four decades of weapon production and research." What
remains obscured in all these claims and assurances is that the DOE
never has done any "health surveys" on people living around its
weapons plants. It even admits that its assessment of health hazards, presumably based on the above-mentioned "health surveys,"
were based on computer simulations and not epidemiological
studies.

   This raises a very real question as to which is more of a safety risk
to the American public—the plants themselves or the government
misrepresentations. At least if people are warned of a danger, as
they were by the European governments after Chernobyl, they can
take some precautionary measures. But the U.S. government has
up until very recently consistently maintained that there were no
health threats from its weapons plants. It insisted that all radioactive materials were safely locked away in sealed, leak-proof containers. Now, it is admitting a threat but denying any harm. But
one look at the events of December 1988 casts serious doubts
around any assertion that no harm has been done, particularly
since many of the practices outlined there have been going on for
many years. Following are summarized findings of the report:

   •**Feed Materials Production Center**, Fernald, Ohio—Two silos
are filled with 390,000 cubic feet of radioactive wastes that emit
harmful radon gas. The plant has also released more than 300,000
pounds of uranium oxide and has contaminated drinking water in
the area.

   •**The Hanford Reservation**, near Richland, Washington—From
the time the plant opened in 1944 until the early 1980s, liquid toxic
and radioactive wastes were dumped in nearby trenches. The

wastes have contaminated large underground reservoirs used for drinking water and irrigation.

•**Idaho National Engineering Laboratory**, near Idaho Falls— The plant, which reprocesses nuclear fuel and operates experimental reactors, has discharged radioactive and toxic wastes into unlined waste lagoons. From there the wastes have leached into the Snake River Aquifer, a giant underground reservoir.

•**The Kansas City Plant**, Kansas City, Missouri—The plant, which produces electronic components for weapons, has released 240 tons of toxic chemicals into the atmosphere. The plant also contaminated soil and sewer lines with PCBs by discharging the cancer-causing chemicals into pits.

•**Lawrence Livermore National Laboratory**, Livermore, California—One of the principal national weapon laboratories, the laboratory has leached toxic contamination from other operations into the groundwater. The contamination at this site is expected to migrate to nearby residential communities dependent on groundwater.

•**Los Alamos National Laboratory**, Los Alamos, New Mexico— Test firings of high explosives have contaminated the test area with uranium as well as other radioactive particles and chemicals.

•**Mound Facility**, Miamisburg, Ohio—This site, which manufactures high explosives and plutonium components for satellites, utilizes waste pits believed to be leaching toxic chemicals into soil. Officials are concerned about plutonium seeping into groundwater.

•**Nevada Test Site**, near Las Vegas—This site is where nuclear weapons were tested above ground until 1961 and below ground since then. About 75 square miles of the site are thoroughly contaminated with radioactive materials from these tests, including plutonium, cesium, and strontium.

•**Pantex Facility**, Amarillo, Texas—A river of toxic solvents from the plant, the final assembly point for nuclear warheads, was discharged into a giant unlined waste pit from 1954 to 1980. The thousands of gallons of chemicals are believed to be leaking into the Ogallala Aquifer, the primary drinking–water source for Amarillo.

•**Pinellas Plant**, Largo, Florida—-Underground storage tanks at this plant, which produces mechanical components for weapons, are believed to be leaking toxic chemical compounds.

•**Portsmouth Uranium Enrichment Complex**, Piketon, Ohio— Roughly 36 pounds of cancer-causing and toxic chromium used to process uranium for reactor fuel is being released into the atmosphere daily through the plant's cooling towers.

• **Rocky Flats Plant**, near Golden, Colorado—The plant, which processes plutonium for weapons, is leaking volatile, carcinogenic organic chemicals into underground water north of Denver. Soil all around the site is contaminated with dangerously elevated levels of plutonium.

• **Sandia National Laboratories**, Albuquerque, New Mexico, Livermore, California, and north of Las Vegas—Miles of desert near the Nevada site are contaminated with hazardous materials used at the laboratories since the early 1940s. Underground storage tanks, long used to store potentially cancer-causing chemicals, have leaked. Open lagoons are receiving contaminated wastes and may be leaking.

• **Savannah River Plant**, near Aiken, South Carolina—The plant, which produces radioactive tritium and plutonium for nuclear weapons, has released millions of curies of tritium gas into the atmosphere from accidents. A major aquifer in the area has been contaminated with solvents.

• **Y-12 Plant**, Oak Ridge, Tennessee—The plant, which fabricates weapon components, has polluted local streams with mercury. A pond at the plant containing arsenic, boron, and sulfate is leaking into surface streams. Crops and livestock on nearby farms may be contaminated by poisonous releases into the atmosphere.

With contamination such as that taking place at the nation's nuclear plants, it is difficult to understand how the federal government can continue to maintain that people living around the plants have suffered no damage. Around the Fernald plant, for example, Charles Zinser and his family had rented a vegetable garden a mile and a half from the plant. In April 1986, Zinser's eight-year-old son Samuel was diagnosed as having a form of leukemia. Three months later, a bone tumor was found on his three-year-old son's leg, and eventually the lower portion of the leg had to be amputated. Doctors treating the Zinser children said that the chance of such occurrences being mere coincidences were extremely low. Zinser thought about the vegetable garden and later had the soil tested. It proved to be contaminated with uranium. The Zinsers and other Fernald area residents are now suing National Lead of Ohio, Inc., the company that managed the site until 1986. While they may eventually receive some financial compensation, the question still remains why the federal government purposely withheld information that could have obviated both the health and property damage in the first place.

Particularly at risk are the more than 600,000 workers who have been employed at the various nuclear facilities since the inception of the Manhattan Project. Little by little reports of worker contamination due to inadequate safety precautions and monitoring systems are leaking out. The DOE admits that workers are poorly monitored and often leave plants with radioactive contamination. Workers then track the radioactivity back to their homes, where it can accumulate and endanger the health of their families. There is a particular irony about the problems surrounding worker safety at these plants. While the employees know that their jobs are more dangerous than usual, they are also, on the whole, a highly patriotic group that views it almost as a sacrilege to question the intention of the government for which it works. Many of them, or their parents before them, were attracted to work at the facilities by government employment drives in the 1950s launched with slogans like "Build your future with atomic energy." Instead of receiving good faith in return for their loyalty and trust, there is ample evidence that the workers at America's nuclear facilities have been the ignored consequences of a nightmarish experiment. At almost every plant, there are stories of men and women who died too young of brain tumors or leukemia.

The safety record of the nuclear industry is grim, and there is little reason to believe that the situation is going to change in the future. Some 10 years after the 1979 accident at Three Mile Island, 80 percent of the nuclear reactors across the country have failed to implement basic safety precautions required under federal regulation after the partial meltdown of the Pennsylvania reactor. This sorry state of affairs prompted Representative Edward J. Markey of Massachusetts to conclude that "we may be as vulnerable to a meltdown in the twenty-first Century as we were in 1979." Part of the problem in completing the safety changes stems from the exorbitant costs involved, which average $50 million per plant. What this in turn suggests is that while it may be theoretically possible to operate a nuclear–power plant safely, realistically this is not going to happen, given the current costs involved. Moreover, points out Representative Markey, "Nuclear power isn't horseshoes. The difference between coming close and getting the job done can mean the difference between safety and catastrophe."

Myth No. 3: Nuclear Is Clean
Commercial nuclear companies are now nearing the fifteenth year of an industry wide slump. No new plants have been ordered in a decade. The last order for a nuclear–power plant that was not

canceled was made in 1975. Few of the plants that are operating are making any money. Cost-overruns seem a matter of course. In 1986, New Hampshire's Seabrook nuclear–power plant was completed at a cost of $4.5 billion. It was five years behind schedule and four times the original cost. Facts and figures such as these make the survival of the industry tenuous at best. But as the world reels under the mounting pressure to do something about the environmental crises that threaten the very existence of the planet, nuclear proponents are hopeful that these crises will provide the industry with a sorely needed revival. They are basing their hope on being able to convince the public that, unlike fossil fuels, nuclear energy is a "clean" energy.

This new promotional concept, coming at the same time as the estimates of the billions of dollars required to clean up after our nuclear plants, may seem ludicrous to some, but the nuclear industry is nevertheless taking it very seriously. The major new selling point is that unlike the burning of fossil fuels, nuclear energy does not emit carbon dioxide, sulfur, nitrous oxides, and other chemical compounds that are believed to be contributing to such environmental problems as the greenhouse effect and acid rain. While it is true that nuclear power does not emit "greenhouse gases," the promotion of this energy as a miracle cure for our environmental woes conveniently omits the discussion of a number of key factors.

First, simply because it does not emit the specific substances that are given off by the burning of fossil fuel hardly makes nuclear a clean energy. Nor, when you consider the radioactive waste that atomic energy does produce, is it easy to even argue that it is cleaner than fossil fuels. Another factor that is rarely considered in the debate is that at least 50 percent of the pollution generated by fossil fuels comes from motor vehicles—cars, buses, and trucks. So, until the time that we have atomically powered automobiles, nuclear will have only a limited impact on our primary source of air pollution. Furthermore, to the extent that large-scale development of nuclear energy serves as a disincentive to conservation, nuclear energy may even contribute to pollution rather than alleviate it. In short, while nuclear may indeed provide a short-term, stop gap solution to some environmental issues, it is hardly the ideal solution, particularly since the problems associated with its use— waste disposal, environmental contamination, health risks, and threats of nuclear weapon proliferation through the illicit conveyance of plutonium—are at least as serious, if not more so, than the problems nuclear is supposed to solve.

Situations belying the "nuclear is clean" myth are found almost everywhere that aspects of the nuclear industry can be spotted. Virtually everything that is used in nuclear plants becomes contaminated and must be disposed of in dump sites. The $700 million Waste Isolation Pilot Plant, 26 miles east of Carlsbad, New Mexico, intended to be the nation's first permanent repository for nuclear waste, was authorized in 1979 for the storage of what is called "transuranic" waste. This is not the unwanted bombs, the used fuel rods, or other exceptionally dangerous waste one normally associates with nuclear power, but instead, everyday things like rags, rubber gloves, and laboratory coats that become contaminated during the normal course of working at a nuclear facility. Even the use of radioactive materials for medical purposes cause waste–disposal problems. In one instance, a citizen's group called the Radioactive Waste Campaign (RWC) discovered that the "repository for the world's largest commercial concentration of radium-226" was found in a densely populated area of Queens, New York. According to the RWC, Geiger counter readings taken on the exterior of the building, occupied by the Radium Chemical Company since 1954, measured 80 millirems per hour, some 40 times the federally accepted standard. The building was used not for nuclear power or some aspect of weapons manufacture, but for the production and servicing of radiation equipment for the treatment of cancer patients.

Concerning by-products of bomb production itself, ever since the detonation of the first atomic bomb in 1942, scientists have been aware that nuclear waste could turn out to be the biggest single problem facing the nation in the future. Up until the seventies the problem went largely unaddressed. Waste was dumped into storage tanks and even into unlined pits. The result has been massive contamination of the ground and aquifers surrounding all of the country's bomb-making facilities. Michio Kaku, professor of nuclear physics at the City University of New York, discusses some of the problems that this "clean" energy is creating with the nation's water supply:

"At the Hanford Reservation site a half a million gallons of liquified toxic waste have leached into the soil. *Half a million gallons!* That is an incredible quantity of high-level waste. It has now reached the water table and is now contaminating the Columbia River. The Snake River aquifers underneath the Idaho Falls site are also now known to be polluted, as is the aquifer under the Pantex site near Amarillo. This type of pollution is particularly serious.

Once aquifers are polluted, it takes hundreds of years for them to decontaminate because water does not move very quickly in these underground reservoirs. There is a real irony here. In the Bible it says that thou shalt not contaminate the wells of thy enemy. However, now we find ourselves in this strange situation where our own government is contaminating the wells, not of its enemy, but of its own citizens."

The pollution of our nation's aquifers by radiation is compounded by the failure of most municipalities to test their water supplies for radioactivity. "Around New York City," relates Kaku, "there was a flash in the news that plutonium had shown up in the city's water supply. But again, the 'authorities' were investigating it, and the story just died. It is tragedy that the departments of health in most cities do not have the scientific equipment that can calculate levels of radioactivity in drinking water."

In the seventies efforts were made to find some more permanent and safe disposal facilities. One hundred million dollars was spent to excavate salt flats in Kansas, but the site had to be abandoned when engineers found that the ground had been punched full of drilling holes for oil and gas. The latest attempt to find a storage site is the New Mexico Waste Isolation Pilot Plant, but after spending even more millions of dollars, federal officials have had to indefinitely postpone its use because there were doubts as to its safety.

The current situation is beginning to suspiciously resemble the trek of the garbage barge from Islip, Long Island, that traveled the world looking for a place to dump its cargo of waste. Increasingly states and citizens are objecting to nuclear waste being dumped, or even transported, through their backyards. In October 1988, the governor of Idaho, Cecil D. Andrus, blocked a shipment of radioactive waste from the Rocky Flats plant in Colorado from being transported through his state to the Idaho National Engineering Laboratory for storage. Governor Andrus, in rejecting the waste, said that it belonged to Colorado and that that state should take it back. Not surprisingly, officials from Colorado were likewise denying responsibility for the cargo.

Nuclear waste is so toxic that even transporting it can be extremely dangerous. Moreover, because of the strategic importance of plutonium, which can be manufactured from wastes generated at nuclear–power plants, transportation can present serious risks to international peace and security. When President Reagan proposed amending a treaty with Japan in order to provide for shipment of plutonium by sea instead of by air, as the treaty had previously required, a number of Congress members expressed grave concerns about the risks of the cargo being hijacked during the long voyage and the costs involved in providing an adequate navy escort to protect it, which the Pentagon estimated at $2.8 million per voyage.

Nuclear technology is causing some of the most serious environmental contamination problems we now face. Unlike any of the other threats to our environment, nuclear power, no matter what we do, will remain here to haunt us for thousands of years. Plutonium-239, for instance, takes nearly a quarter of a million years just to decay into another radioactive substance, uranium-235. We have seen that the costs of simply cleaning up after our currently operating nuclear facilities can range in the billions and possibly trillions of dollars. The costs are so staggering that there remains the very real possibility that many sources of contamination will be abandoned.

Nuclear–Energy Politics

No issue in America today is more politically charged than that of nuclear technology. Having commenced under the veil of secrecy of the Manhattan Project, which was set up under President Franklin Roosevelt to manufacture and develop the atomic bomb, the roots of the nuclear industry have been firmly planted in our society without most Americans even realizing what was taking place. Its scope ranges from bomb-making, to energy production, to medicine, and most recently there are even massive campaigns by the nuclear industry calling for the irradiation of the nation's food supply. The interests involved include many of the most influential and respected sectors of society—the defense, energy, and medical establishments, government, universities, and American's most powerful corporations.

In October 1988, a series of revelations jolted the nation into realizing just how "unthinkable" some of the practices at the nation's weapon-making facilities really were. Workers went on strike at the Feed Materials Production Center in Fernald, Ohio, over pay and safety conditions. This focused attention on practices that had been taking place for years at the Fernald plant, particularly the admittedly deliberate release of uranium dust into the air. Early in October the plutonium processing site at Rocky Flats, Colorado, was closed after unprotected workers became contaminated when they walked into an area where other workers were cleaning up a plutonium spill. These two situations at these two plants served as a backdrop to revelations concerning serious problems at the Savannah River plutonium-processing facility in South Carolina.

Attention was first drawn to that plant in late September 1988, following a congressional hearing in which Senator John Glenn of Ohio and Representative Mike Synar of Oklahoma released a memorandum from E. I. Du Pont de Nemours & Company outlining more than 30 significant reactor accidents that had taken place between 1957 and 1985. Du Pont built the Savannah River plant in the early 1950s and managed it until April 1989 when Westinghouse took over. The memorandum was apparently written with the intent to show progress in safety, the argument being that serious mishaps were becoming less frequent. But Congress was not quite as sanguine about the contents of the memo, particularly the report that an error by a Du Pont foreman almost turned a leak of reactor coolant into an accident that would have destroyed the reactor. Following the congressional study of the memorandum, the Energy Department did release a study telling of a wide number of safety problems and mishaps, ranging from equipment failures to worker contamination. The study estimated that over a 20-year period, the plants had to shut down on the average of 9 to 12 times, nearly twice the rate for civilian power reactors. Nuclear experts commenting on the report said that if similar occurrences had taken place at civilian reactors, they would have been shut down. Almost as shocking as the events themselves was the manner in which near-disasters had been kept secret, some for as long as 31 years. While the DOE initially denied any knowledge of the events reported in the study, it later admitted that either it or its predecessor organization, the Atomic Energy Commission (AEC), were responsible for keeping the matter hushed up. On the other hand, former AEC officials commented that the situation had been kept secret by local managers, presumably because releasing information of internal problems with bomb production presented a risk to national security. John S. Herrington, then-secretary of the DOE, made comments following the release of the report suggesting that the veil of national security had protected plant managers and permitted them to do things that would not be possible elsewhere. "Du Pont had the idea there was nobody in the universe who could tell them anything," he said. Herrington also noted that the DOE had worked so closely with Du Pont employees, it was difficult to differentiate the two. While the Nuclear Regulatory Commission, the federal agency responsible for supervising the civilian nuclear industry, is "toothless," he said, "at least they've got a watchdog."

Shortly after the Energy Department's report was released, further details about operations at Savannah River were made public by the Environmental Policy Institute in Washington, a nonprofit group investigating the DOE's weapons facilities. Via the Freedom of Information Act, the group had received technical reports setting forth a series of mishaps at the plant, including fires, equipment failures, a flood of contaminated water, and a plutonium leak that narrowly missed setting off a nuclear reaction in the plant. Discrepancies between what Du Pont said about some of the incidents and subsequent comments on the severity of the accidents by government officials suggests serious attitude problems on the part of Du Pont. One of the instances involved the flooding of a plutonium-processing area. The water leaked into an adjoining office where plant employees were stationed. William Durant, a senior researcher at Du Pont, interviewed about the accident said that it was not significant. "It was plutonium-contaminated water," he said. "As long as you don't get it inside your body, it won't hurt you." However, Deputy Assistant Secretary of Energy Richard Starostecki was not quite so cavalier about the leak. "Any time you have plutonium in an uncontrolled fashion it's serious," he said. "This was in water, which evaporates, and during the cleanup probably becomes airborne. People were probably breathing it in."

When an accident on November 6, 1978, caused the release of radioactive ruthenium to be dispersed throughout the plant and beyond the plant boundaries, DuPont claimed that the amount of radiation exposure was less than one quarter the amount of normal background radiation. But Starostecki and other health officials noted that the danger from ruthenium was not from direct exposure, but from inhalation, in which case it could be very dangerous.

Du Pont officials insisted that the DOE's criticism of its management of the Savannah River site was merely political and had no basis in fact. The chairman and chief executive officer, Richard E. Heckert, complained that Du Pont was getting a "bum rap" and said, "It's a political thing, not an equipment problem whatsever." In his opinion, the DOE was exaggerating problems in order to get money for a new reactor and was trying to make Westinghouse, the company scheduled to take over from Du Pont in April 1989, appear in a more favorable light. There is undoubtedly plenty of truth in Du Pont's claims. After 40 years of secrecy and the deliberate release of massive amounts of highly toxic materials, it is hard to believe that the DOE's sudden candor stems from a deepfelt concern about the welfare of the public. On the other hand, Du Pont's cry of "politics" to defend its performance record, when it was politics that allowed it to perform so abysmally in the first place, comes very close to the ultimate irony. It does, however, show even more clearly the degree of political maneuvering that takes place in the nuclear arena.

Dr. John Gofman, an author and professor of medical physics at the University of California, feels very strongly about the role politics is playing in the nuclear debate, particularly the manner in which it is affecting scientific research. "The vested interest in having people believe that there is little hazard in radiation is enormous," he says. "It involves medical, dental, industrial, nuclear activities amounting to billions of dollars, and they put up a very good countercampaign to work such as mine that exposes the risks inherent in radiation. In fact, most recently the nuclear supporters have mounted a campaign to tell you that taking some extra radiation is good for you. They have given it a scientific name—hormesis—and are now telling us that everyone needs a little jolt of radiation to get their immune system working well. Or if you look at reports from the Department of Energy and some of its scientists they will tell you that the net effects out of Chernobyl might be close to zero extra cancers. My estimate for Chernobyl worldwide is a million extra cancers. So don't underestimate the countercampaign. You can put out some truths. You can try to get people from both sides to come forward, but remember there will be a countercampaign that would make your head swim. Ninety to ninety-five percent of the funding of all medical research comes from the government of the United States and that government is a chief sponsor of activities that irradiate people. You are not going to find many scientists who are going to get 'the wrong answer' about radiation. Your funding does not survive long if you get the wrong answer. So when somebody tells you that there is a 'consensus of scientists' who find no problem with radiation, you bet there is a consensus!"

Low-Level Radiation

A growing number of scientists, in analyzing radiation data and health statistics, are putting forth hypotheses that are in direct contradiction to the conclusions reached by the early studies performed on the Hiroshima data, namely that even very small levels of radiation given over a long period of time, as opposed to the one time massive exposure of the Hiroshima experience, may be presenting risks to health that were previously ignored.

Dr. Carl Morgan is the former head of the Oak Ridge National Laboratory and was also a member of the International Commission on Radiological Protection where he chaired the Internal Dose Committee. After many years of research and experience on radiation, Dr. Morgan is now coming to the conclusion that has long been heresy among government scientists. "All radiation is harmful," he says. "There is no safe level of radiation exposure any more than there is a safe time to go without your seat belt. One should avoid all exposure to ionizing radiation whenever possible. The slope of the curve is greater at low doses than at high doses. You get more cancers per unit dose at a low dose than you do at high dose. This is a matter of great concern to all of us."

Dr. Ernest Sternglass, professor emeritus of radiological physics at the University of Pittsburgh, has done an extensive amount of research into the effects of low-level doses of radiation on human health. He gives a sharp overview of the problem:

"One of the things that really misled scientists for a long time about the effects of low-level radiation was the belief that low-level radiation such as ordinary chest and dental Xrays were relatively benign. Consequently, we really did not expect that very tiny traces of fallout from bomb tests or accidents like Chernobyl could have such large effects on human beings based on all the knowledge we had on medical Xrays. This is why it has been so difficult to have the scientific community examine this. All our previous experience seemed to show that for average healthy adults, chest Xrays do not have a significant impact on human health. But now after Chernobyl, it now looks that somewhere between twenty thousand to forty thousand people died in excess of normal expectations in the United States during the summer 1986 following the accident. These deaths simply cannot be reconciled with the theories we had concerning radiation following the studies on Hiroshima. This is why we have to completely rethink what is going on and reevaluate the biological problems that are involved with fallout in the environment."

Dr. Sternglass says that another reason this issue of low-level radiation has gone ignored for so long is that attention in the scientific community has been focused only on the "direct hit" effect of radiation on the DNA of the cell. "We were also misled," he says, "by experiments showing that it took thousands of rads—you get about one tenth of a rad a year from normal background radiation—to rupture the cell membrane. All our radiation standards have been set up only to take into account these types of effects, and hence attention has only been focused on large levels of radiation and the damage it does to the genetic material. But, it turns out that there is a completely different biological mechanism that causes damage. In this case, the radiation does not act so much on the genes directly, but instead produces highly reactive chemical species called 'free radicals.' These are unstable forms of ordinary elements like oxygen. For example, when an oxygen molecule has a negatively charged electron attached to it, as produced by radiation entering the cell, then this molecule goes through the cell membrane and literally 'unzips' it, causing the cell to die. This is much more efficient at low doses than at high doses. Additionally, if the radiation is given gradually over a period of days or weeks, which is the case when we ingest or inhale radioactivity, and it sits in our bones and irradiates the bone marrow, it can be thousands of times more damaging to the immune system than we had calculated according to the genetic effect."

According to Dr. Sternglass, the single intense exposure that you get from something like medical Xrays are like tiny little bullets hitting the DNA in the center of the cell. This type of damage, he says, is quite efficiently repaired by enzymes that go up and down the strands of the DNA. So it takes quite a large dose of this direct bulletlike action to damage the cell's DNA. But he believes that the damage caused by free radicals produced by low levels of radiation not only produces a different effect from the "direct hit" damage to the DNA, but also is more damaging at lower levels of exposure. "If too many of these free radicals are created all at once, as in a very short burst, then they become ineffective because they bump into one another."

As to why we are not being given the information that we need concerning the effects of this low–dosage effect of radiation, Dr. Sternglass believes that the release of such information would be devastating for the military uses of nuclear weapons.

He estimates that about 10,000 to 20,000 excess deaths occur each year from such things as medical Xrays, and about five times as many from the leakages and normal small emissions of nuclear reactors. One of the things that Dr. Sternglass finds particularly disturbing is the contamination of the food and water supply caused by even small leaks from nuclear facilities. For instance, radioactive iodine-131 when deposited on grass will be eaten by cows and thereby finds its way into the milk supply. "When a pregnant woman drinks the milk," says Dr. Sternglass, "most of the iodine goes right to the thyroid of the developing fetus. As milk is highly perishable, it is transported, often many miles, within a matter of hours or a day, during which time the radioactive iodine remains very active. So in only a few days, the radioactive iodine from the nuclear plant has found its way to the developing fetus. This iodine inhibits the proper function of the fetal thyroid causing the baby to be born underweight or prematurely. It gives this epidemic of very low birth-weight babies that we have been seeing over the past forty years."

The American public was led to believe that the accidents at Three Mile Island or Chernobyl had no adverse health effects on people in the United States. Jay Gould is a statistician and a fellow at the Institute for Policy Studies in Washington, D.C. For the past several years, he has investigated the mortality effect of nuclear accidents and has found that, particularly in the period since 1975, there have been "statistically significant increases" in total mortality in areas close to *all* nuclear reactor sites in the United States. This result, says Gould, was "quite unexpected" since the scientific community and the government had been adamant about there being no adverse effects from nuclear–power plants.

In the case of Chernobyl, there was an increase in mortality in the United States during the summer months of May, June, July, and August that made that period responsible for the highest percentage increase in mortality that has been documented in 86 years. The probability that such an increase was due to mere chance is less than 1 in 10 million.

Of the total deaths, the majority occurred in the age groups of 65 and over. Fetal deaths also increased enormously and were not included in the mortality rate statistics. This almost completely correlated if you examined the levels of radioactive iodine found in milk during those summer months. The largest decrease of live births was found in the Pacific areas, which had a large amount of rainfall that caused heavy amounts of radioactive deposits to enter the food chain. This region increased its summertime mortality rate by 5 percent, as compared to places like Texas, which has very little rainfall and no increase in mortality. This very strongly suggests that the rate of mortality is related to the amounts of radioactivity that was deposited in given areas.

Both Gould and Dr. Sternglass cite estimates by Dr. Romley Burtelle that the total number of excess deaths that could be attributed to low-level radiation since 1945 could be anywhere between16 million and 30 million. Dr. Sternglass says that his calculations and those of others researching the area confirm the findings of Dr. Burtelle. "We all agree," says Dr. Sternglass, "that the worldwide testing of the nuclear bomb by itself must have resulted in excess deaths [above normal expectation due to all causes] something in the order of thirty million. It is almost as if we had had another World War II, but a silent one."

Concerning the number of cancer deaths are occurring each year due to low-level radiation, Dr. John Gofman of the University of California estimates that at least 15 percent, or about 75,000, of the nation's cancer deaths are caused by radiation—including medical and background radiation. He adds, however, that if medical radiation was improved so that it eliminated defective equipment and gave only the dosage necessary to get a proper diagnostic picture, it would be easy to cut the dose on the average of one third. "This measure alone," he says, "could prevent about fifty thousand cancers each year, which in one generation of thirty years would be a million and a half cancers prevented."

Chapter 10
Electromagnetic Pollution

Electronic smog is all around us—it emanates from our color–television sets, video display terminals of computers, the power lines zigzagging the landscape, as well as from all electrical appliances. Everyone is affected by it. So, the issue is not whether electromagnetic radiation exists, but whether it is adversely affecting our health. If so, then the question arises as to how and to what extent we are being affected, and what we can do to minimize any damage in the future.

Electricity is an integral component of today's society. If it turns out that all of our electrical conveniences are capable of causing health problems, the solution is hardly going to be an instant cessation of using this form of power to fuel our energy-hungry nation. Many of the experts discussing the effects of electromagnetic radiation today say that there is no need to be concerned because there is not enough evidence of harm. So, as in instances like DDT, asbestos, and other substances suspected of causing harm, we have to ask ourselves when and at what point do we decide that we have enough information? For 40 years, the cigarette lobby maintained (and continues to maintain) that there is not enough evidence to prove that cigarette smoking is truly harmful. That is after thousands of scientific studies proving its toxicity.

Outlining the Problem
Electromagnetic energy is measured by its frequency—the number of waves or cycles passing a point in a given period of time. The standard unit of measurement is the hertz (Hz) which is defined as one cycle per second. One kilohertz (kHz) is 1,000 cycles per second, 1 megahertz equaling 1 million hertz. We have long known that radiation in the higher zones of frequency like Xrays and

gamma rays can cause cells to mutate and, with large exposures, to die. The energy we will be talking about in this chapter concerns radiation falling within the lower–frequency range and which now is suspected of causing significant adverse health effects.

With the myriad of health and environmental problems now facing the nation, electricity has long seemed to be our one bastion of safe and clean technology. Sure, the coal, gas, or nuclear power used to generate it is causing pollution, but if we could find some way to generate electricity without the greenhouse gases, the oil spills, and nuclear waste, all our problems would dissolve. The problem is not with the electricity itself, right? So it seemed until 1979 when researchers completed a study that set all the assumptions concerning the benign nature of electricity on their head. Children in Denver living close to high–voltage lines were dying of leukemia and other cancers at an abnormally high rate.

The work of researchers Nancy Wertheimer and Ed Leeper represented the first epidemiological study suggesting a link between cancer in children and the electromagnetic fields surrounding the high-voltage power lines. It was immediately dismissed by critics as being flawed—because of their low budget, the two researchers estimated instead of actually measuring the voltage surrounding the power lines. The public, however, was not so easily dissuaded, and concern among citizens continued to grow. Eight years later, a second group of scientists endeavored to set the record straight. But rather than disproving the Wertheimer and Leeper study, the new study, headed by David Savitz of the University of North Carolina, uncovered similar findings. Also conducted in the Denver area, the Savitz study concluded that children subjected to the electromagnetic fields generated from power lines near their homes were 1.7 times more likely to develop cancer than children living in lower field areas. For leukemia and other soft-tissue cancers, the risk was even greater. Recently testifying before Congress, Savitz said that the results of his study were suggestive rather than conclusive. But he added if the results are confirmed by other researchers, it could mean that a large proportion of childhood cancers are related to the electromagnetic fields generated around power lines.

While the Wertheimer and Leeper study was a lone voice on the subject of electromagnetic pollution and its effects on human health, the Savitz study came at a time when scientists were beginning to discover more about the atomic behavior of cells, much of which depends on electronic signals. Researchers have proven that low-frequency electromagnetic fields can disrupt the body's natural immune system, modify the production of hormones, and help promote the growth of tumors. Some now believe that physical events, rather than chemical ones, determine biological functions.

Scientists have seen cells detect an electromagnetic field as small as a ten-millionth of a volt per inch, and are concerned about cellular response to fields as much as a thousand times stronger. What is the reaction to the field of one volt per inch—the exposure induced by holding a cellular telephone?

Scientists are far from reaching an agreement as to the mechanisms by which electromagnetic fields may be causing health problems, but many are convinced that there is damage occurring. Jerry L. Phillips, director of biomedical research at the Cancer Therapy and Research Center in San Antonio, Texas, is one of the leading researchers in the area of electromagnetic fields and their relationship to cancer and irregular cell growth. "I have concerns in a number of areas," says Dr. Phillips, "particularly regarding the association between long-term exposure to low-level electromagnetic fields at power–line frequency and an increased incidence of cancer in both adults and children. I am also worried that similar exposures may be associated with other health hazards in the areas of behavior, reproduction, immune response. While the evidence is not conclusive—and probably never will be—scientific information available now is highly suggestive that there are problems." He thinks that while electromagnetic fields may not be carcinogenic in themselves, they may work to promote cancer by weakening the body's immune response.

Unlike some other scientists researching the area of electromagnetic radiation, Phillips's work has also led him to believe this type of energy can actually cause damage to the cell's DNA. There are some recent studies bearing on the role electromagnetic–field exposure may have on the initiation of cancer. There is one study by a group of German researchers and another by an Egyptian group in the scientific literature. Both found an increase in damage to a cell's DNA because of exposure to power–line–frequency electric fields. There is a third study reported by a group of Swedish researchers recently at the annual meeting of the Bioelectromagnetic Society. They found that exposure of cells to electromagnetic fields of power–line frequency produced an increase in damage to those cells.

One factor that has led many researchers to discount the biological effects of electromagnetic radiation is that the correlation between dosage and damage seems to be the reverse of what is normally expected. Scientists are familiar with an increasing dosage of a harmful substance causing an increase in damage. With electromagnetic radiation, however, experiments are showing that damage may increase with exposure to *weaker* electromagnetic fields. W. Ross Adey, associate chief of research and development at the Veterans Administration Medical Center in Loma Linda, California, theorizes that damage resulting from exposure to electromagnetic radiation may not depend on the dose (the strength of the field), but instead on what he calls *windows*. While a single exposure to strong fields may have little effect on an organism, the window theory takes into account the frequency and strength of the field as well as the period of exposure time. Adey discovered in his research that exposure of cells to a particular low-frequency electromagnetic field caused the movement of calcium from the cells. At higher frequencies, the movement stopped. Calcium is important in many cellular functions, including the activity of the body's white-blood cells. The proper functioning of these cells is crucial to an effective immune system.

Synergy and Accounting for the Differences Between Us

There are two factors commonly overlooked or discounted by many scientists today. One is that of the synergistic effect of environmental pollutants in which the effect of two agents together becomes greater than the sum of the effects of the two agents taken individually. Scientists and industry officials are fond of defending the use of pesticides, for example, by saying that residues are so small that they could not possibly cause damage, or that there are numerous naturally occurring carcinogens, so, the reasoning goes, what difference should pesticide residues make? What these arguments ignore is that scientists to date have very little idea of the interactions of various chemicals and environmental toxins. In the area of electromagnetic fields, researchers are already discovering that there may very well be significant synergistic effects with other agents. "There must be a concern for the interaction of agents in the environment," says Phillips. "In one study, German researchers looked at the interaction of electric–field exposure and chemical treatment to cells. They found an increase in genetic damage to those cells when they were treated with both DNA-damaging chemicals and the power–line–frequency electromagnetic fields."

The second factor receiving little attention from science and medicine has to to with the different tolerance or susceptibility levels of each person. A person with AIDS, for instance, will be much more susceptible to all sorts of bacteria and environmental pollutants because his or her immune system is powerless to defend against these agents. Likewise, children, pregnant and lactating women, and people under inordinate stress often show less resistance to toxins in their environment. "We need to recognize," says Phillips, "that susceptibility to different agents in the environment varies with the individual. We all recognize the hazards associated with smoking, and yet not every smoker will develop lung cancer. So there is certainly a difference in genetic makeup that allows for a difference in sensitivity to various things in the environment."

Both of these factors need considerably more attention from scientists and researchers than they have received to date. While many scientists are still arguing that there is little or no risk from electromagnetic fields, evidence is already accumulating that children and pregnant women are particularly susceptible to this type of pollution.

Military Production of Electronic Smog

Another large source of electromagnetic pollution is emitted from military testings. One such test coming under significant criticism is the electromagnetic–pulse test, conducted in Maryland, Virginia, Alabama, New Mexico, and California. The Foundation on Economic Trends, a nonprofit environmental group, brought suit in federal court against the Pentagon seeking a halt to electromagnetic pulse tests, and in May 1988, an order was issued enjoining the Pentagon from further testing pending the study of health and environmental effects resulting from high-energy experiments. The foundation claimed that the Pentagon testing not only presented risks to human health but also threatened to disrupt functions at nuclear–power plants, telephone facilities, computers, and automobiles. "This is the first time that the Department of Defense by court order has been forced to take part of its high-tech nuclear program off-line," said Jeremy Rifkin, president of the foundation. "This is the first recognition that electronic pollution is an environmental issue."

Electromagnetic pulse (EMP), a form of nonionizing radiation, is the primary type of electronic pollution engaged in by the military. EMP is one of the aftershock features of nuclear detonation, and the military's main purpose in conducting EMP testing is to determine how a nuclear explosion will affect its sensitive electronic equipment. The military has pursued EMP testing since the 1950s on two fronts. One is to attempt to make military hardware and communications systems invulnerable to nuclear–explosion EMP. The second use is part of the Star Wars program.

At present, there are not many EMP facilities operating. There is one in White Sands, Utah, and there is one off the coast of North Carolina being run by the navy. Most of the others were closed following the Foundation for Environmental Trends lawsuit because it was impossible for the military to produce adequate environmental documentation as to the safety of EMP facilities.

According to Smith, even the purpose of EMP simulators needs to be reexamined. "There is no good effect that is going to come out of the hardening of military devices," he says. "It points to the issue of nuclear survivability, which has been totally discredited by the scientific community. It can be seen as nothing but war promoting by the Soviets, who have got see our intense research into EMP as research into first-strike capacity. Even though that is not the stated purpose, that is the obvious outgrowth of it. There are any number of alternatives to actual above-ground testing with EMP simulators. The military at large has had great resistance to laboratory testing of devices or to computer modeling, even though there has never been any credible evidence put forth as to why these would not be suitable alternatives."

Controversy in Eugene, Oregon

In early 1978, University of Oregon staff members began dogging public–health officials in Eugene about a powerful radio signal. The mysterious signal was causing a number of health problems to the residents of the community. Experts began monitoring and confirmed the existence of the signal, describing it as a 4.75 megahertz pulse occurring 1,100 times per second, but remained baffled as to its origin. At some points, it was measured at power intensities of 500,000 watts—ten times the level of the most powerful AM radio station licensed by the Federal Communications Commission (FCC).

An engineer called in by the Oregon State Health Division to track the signal found that more than 10 miles from any known transmitter it was still registering at a power level of about 500,000 watts. The strength corresponded to a 1 watt transmitter 10 feet away or a 500,000 watt transmitter 10 miles away. The signal didn't diminish as researchers changed their location, and no known transmitters of that size exist in the vicinity—they concluded that the signal came from power lines that acted as a huge antenna and produced the radio wave.

Despite the strength of the signal and the involvement of both local scientists and federal officials, the source of the pulse was never conclusively determined. The FCC explained the signal as datalink operations by the navy. "These are authorized, legitimate navy operations," said FCC spokesman Richard Smith. "There is nothing unusual about that."

The EPA, rushing in to do its own monitoring, also dismissed Shrock's conclusion and backed the FCC position that the unidentified radio signal was coming from a navy transmitter in Dixon, California. Richard Tell, the EPA physicist conducting the monitoring, assured the citizens of Eugene that the Dixon signal could not possibly cause health effects. In the meantime, however, coinciding with the detection of the signal, people were reporting a number of adverse health effects ranging from ear and throat problems to headaches and insomnia.

Some of the Evidence

Matthew Ryan is the editor of *Wildfire Magazine* and is particularly concerned about the impact of fluorescent lighting and video–display terminals (VDTs) on our health. According to Ryan, much of the work establishing the correlation between biological damage and electromagnetic radiation was conducted by John Ott in the 1950s when he was doing stop-action photographs for the Disney studios. While attempting to film the growth of a bean plant through its full succession, Ott discovered that the plants began to exhibit a number of problems when placed at the end of the fluorescent light tubes he was using. Ott was only able to stop the destruction of these plants, or the inhibition of their growth, by placing a lead shield in between the plants and the ends of the fluorescent tubes. Ott later began to investigate the emissions coming out of color–television sets. This resulted in Congress passing a law restricting these emissions. He was finally able to lessen the effects of the color sets by again placing a lead screen between the TV and the plants. But, says Ryan, "When the video display terminals came out, Ott attempted to replicate the same experiments, and to date, he has been unable to protect plants from the effects of VDTs despite using three lead screens and placing the bean plants ten feet away. Obviously, we're dealing with something of an increased problem here."

One of the evidentiary problems with proving the harmful effects of electromagnetic radiation is that there is no one specific condition caused by this type of pollution. Rather, the results may turn out to be a sort of synergistic weakening of the entire body. "It is not so much a question that a fluorescent light will cause a specific problem," says Ryan, "but that it becomes an additional burden on the vitality or on the immune system of an organism. One of the salient features of the research into the ELF fields (those electromagnetic fields vibrating particularly at thirty to one hundred Hz) is that they consistently interfere with the cues that keep an organism's biological cycles properly timed. This then results in chronic stress and impaired disease resistance. Increased cancer rates and reproductive problems are very common findings in studies using electromagnetic fields, particularly in the ELF range."

Part of the problem with ELF frequencies is that they are not attenuated, so that theoretically, at least, Soviet tests will affect people here on this continent and vice-versa. With respect to earthquake activity, scientists have noted that there appears to be a seven-to-ten–day lag following an underground nuclear test, during which there is a buildup of unexpressed energy in the earth's core which then results in an earthquake a few days later. Dr. James DeMayo points out that the earthquake of December 1988 in Armenia followed a very major Soviet test in that area. We know that seismic activity has increased in California since underground nuclear testing began. Scientists are also predicting not only increased seismic activity, but also increased *force* of seismic activity. They are additionally attributing the underground testing to the very accelerated rates of heat buildup in Mount Rainier and Mount Baker, which has been verified by NASA."

More Debate, More Controversy... Any Solutions?

As with the situation in Eugene, Oregon, the debate about the effects of electromagnetic pollution involves many who say that there is no reason to worry. William Bailey is a scientist who has conducted research for the past nine years on the biological effects of electromagnetic fields. He currently works with Environmental Research Information, a scientific consulting firm. Bailey's position is that there is no scientific evidence indicating that we should issue warnings, but that we do need additional testing.

As to studies suggesting that electromagnetic radiation may lead to a variety of health problems, including an increase in miscarriages in pregnant women, Bailey says that since the 1970s there have been about eight such studies. These studies have *not* found an association between the use of VDTs and increased miscarriages in women.

But researcher Jeff Sorensen says there is a very legitimate concern about the health problems that VDTs may be causing. There are about 30 million of these VDTs in use today, up from about 1 million just 10 years ago. VDTs have become one of the most common sources of nonionizing radiation that people are routinely exposed to. People sit right next to them and stare into them during their workday. This is unusual compared to many of the other sources of radiation. If you are a newspaper copyeditor, for instance, you are sitting inches from the screen all day long everyday.

VDTs produce very low–frequency radiation from a mechanism called the fly-back transformer which is found in all computers. There have been many studies about VDTs and problems with pregnancies and other health disorders. At the *Toronto Star*, United Airlines, and *USA Today*, there have been a number of women working on the terminals who have had problems like birth defects and miscarriages.

There was also a study conducted by the Kaiser-Permanente medical program in Oakland, California, which found that women who used VDTs in the first three months of pregnancy for more than 20 hours per week tended to suffer about twice the number of miscarriages as the other women in the office.

In June 1989, the Office of Technology Assessment (OTA) issued a report concluding that more research is required to determine the potential health effects of electromagnetic fields, particularly on producing nervous system disorders and promoting cancer. While the focus of the OTA study was on high-voltage power lines, researchers also indicated that more attention needs to be devoted to lower–voltage emissions coming from things like household wiring and appliances. According to the study, alternating current of 60 cycles a second, which is used in most households, and other low-frequency electromagnetic fields can interact with individual cells and organs to produce biological changes. The nature of these interactions for public health remain unclear, but there are legitimate reasons for concern.

H. Keith Florig is one of the authors of the OTA report and is affiliated with the Department of Engineering and Public Policy at Carnegie-Mellon University. Like Bailey's assessment of data on electromagnetic research, Florig's analysis of the results of both his studies and the OTA study are equivocal and noncommittal. Also notable is the divergence between Florig's assessment of his own work and that of the final report issued by the OTA.

"In our study," he says, "we looked at three things—transmission lines, distribution lines, and appliances. We reviewed the scientific literature on the biological effects in relationship to the problem of transmission–line siting. We concluded that the scientific evidence seems much too incomplete to provide the kind of risk assessment that are needed to produce regulatory policies. There are several reasons that the science is so confusing and contradictory. First, the health effects, if they exist at all, might be very subtle—they might be beyond the limits of science to see them. The biological–effects mechanisms might be very complicated. We have seen evidence at the cellular level that effects increase with a decreasing field strength. That is backwards from what we are used to. Another reason that the evidence seems confusing and contradictory is that the bio-effects data are often contaminated by false–negative and false–positive results. This is an inevitable result of doing research; you are always going to generate studies that show an effect when really there isn't one, and studies that fail to show an effect when there is one. And finally the theoretical basis for how fields react with cells are in a very early stage of development. The point is that it might be many tens of years before we have a theory about how nonionizing fields produced by powerlines can interact with cells to produce biological effects.

With respect to the Kaiser-Permanente study, the authors of that study found that women who worked with VDTs for more than 20 hours a week experienced both early and late miscarriage rates 80 percent higher than the risk for women who performed similar work without using VDTs. The authors of this study wrote, "Our case–controlled study provides the first evidence based on substantial numbers of pregnant VDT operators suggests that high usage of VDT may increase the risk of miscarriage."

While scientists and utility representatives are fighting it out as to whether electromagnetic fields are causing injury to health, the public has not been as unwilling to start taking preventive measures. According to David Carpenter, a New York State public–health official, the findings of links between electromagnetic fields and the increased risks of cancer "are sufficiently worrisome that we should begin to change the way we wire our homes and not delay for another five to ten years for additional epidemiological studies."

Florida officials have recently drawn up regulations to limit the amount of electromagnetism that new power lines may generate. Six states already limit the intensity of electrical fields around power lines. In California and Florida, utilities are facing lawsuits from school districts fearful of potentially harmful health effects from proposed high-voltage power lines. And in Texas, Houston

Lighting & Power Company appealed a multi-million dollar punitive damages award granted by a Houston jury to the Klein Independent School District. District officials filed suit after the utility strung a 345,000-volt line next to a school complex with 6,500 students.

"People were concerned there were unanswered health questions about that line," says Donald Collins, Klein's superintendent. "They didn't want their children to be subjected to an experiment." A spokesman for the utility said the high-voltage line has been disconnected and is being rerouted far away from the school complex.

In July 1987, New York State health authorities published the results of its Power Line Research Project. At that time, Dr. David Carpenter, a public–health official who headed the study, commented that if the results of the study were correct, as many as 10 to 15 percent of the cases of childhood cancer may result from residential electromagnetic field exposure.

The New York Power Line Research Project was a five-year research project, funded by New York state's utilities, ordered by the state's public service commission and administered by the Department of Health to determine whether or not there were health hazards from exposure to electromagnetic fields. Five million dollars was assessed against the utilities in order to fund the 16 different research projects. The final report of this project was published in July 1987.

There were three major areas of concern in this research program. One was whether electromagnetic fields caused genetic effects or birth defects. In that area, the answer was a fairly clear no. The second area of major concern was that of nervous system and behavioral alterations. In this case, almost half of the research projects we supported were concerned with these problems. In almost every study, there were some positive observations found. Many of the effects of electromagnetic fields are not clearly translated into being hazardous, but there is no question whatsoever that they cause effects on animal behavior and on the nervous system. In a few circumstances, there was reason to be concerned that the effects of the electromagnetic fields were ones that could not be good for you.

The final question was cancer. This was studied in human populations in epidemiological studies, principally the work of David Savitz confirming the original work of Nancy Wertheimer and Ed Leeper, demonstrating that in homes where the magnetic fields were elevated as a result of the power–distribution lines, the incidence of childhood cancer was double as compared with those homes where the magnetic fields were lower. On the basis of these studies, the panel of experts concluded that without question electromagnetic fields can cause biological effects. The areas of concern are principally cance;, secondarily, and at somewhat high field strengths, effects on the brain and behavior.

Prevention and Suggestions for Decreasing Risk

While there is always room for more scientific study, there comes a time when sensible preventive measures have to be taken. Studying the problem is not a preventive measure. Discarding electric blankets is a good preventive measure, and all Americans should do that right now. Staying an arm's length away from your video–display terminal and making sure that your children stay that distance from the television set when they are playing video games or watching TV is a good preventive measure. We could lower significantly exposure to the magnetic fields that appear to be causing the problem by doing just these few things. People who live near high-current wires across the country should be pressuring their town and city health officers to assess with gaussmeters (a very simple device costing about $200) the magnetic fields, find out where they are high, and if they are high, insist that the utilities rewire to get these fields away from people.

According to Jerry Phillips, while electromagnetic fields are all around us, there are only a few areas demanding immediate attention if health risks are to be avoided:

"I am always concerned about high-voltage power lines to residences and schools. That is an issue that needs to be dealt with at an international level. Within the home there are only a couple of appliances that are of any concern whatsoever. Those are the VDT unit and the electric blanket. The VDT because the fields can be substantial at a variety of frequencies, and the blanket because it is right on top of people. With the exception of these two items, exposure to electromagnetic fields from household appliances is spotty at best and does not present a problem for most people."

We have also noted that officials in a number of states have already begun acting to restrict the electric fields surrounding new power–line installations. Florida has gone even further and is drawing up regulations to decrease both electric and magnetic fields. Rather than waiting until definitive proof is in, citizens need to urge their local and state officials to, at a minimum, set standards for all new power–line installations. Testing existing power lines for "hot spots," and requiring rewiring to get these fields away from inhabited areas is also important.

Protecting Yourself from Bad Vibes

In June 1989, the OTA stated that more research needs to be done in the area of electromagnetic radiation and its effects on human health. According to the report issued by the OTA, the type of current most commonly used in American households and other low-frequency electromagnetic fields can interact with individual cells and organs to produce biological changes. The nature of these interactions for public health remain unclear, but there are legitimate reasons for concern.

While such study to be undertaken by Congress is to be encouraged, the American public also needs to ask itself at this time whether study is enough. Study is not a preventive measure and in many cases can merely serve to unnecessarily delay decisive action when it is sorely needed. Moreover, even when the studies are done, there are always those who will simply ignore the results of these studies and deny that they say anything significant.

If anything, learning about the risks involved teaches us how to really be concerned about product safety. That is the responsibility of industry and the scientists who work for industry. Most of these people tend to be conscientious about trying to create a better and safer product. But sometimes responsibility also devolves onto the public to show industry that products need to be improved. We have seen how this works with automobiles. We continue to improve them to this day. If we did not raise concern over certain safety issues, chances are industry would have either not addressed them at all or would have taken much longer to do so.

"We could study this problem for the next twenty years," says Ryan. "We certainly do not know all the mechanisms involved and what is exactly causing the problem. But I don't think that there is any question that we do have a problem." Some of the experts will say, "Oh, well, we haven't proved that electromagnetic fields can cause cancer. Until we do that, there is no need for alarm." But is that really true? There is ample scientific evidence that electromagnetic radiation can induce cancer. Is it not time to shift the burden of proof and err on the side of caution until industry can prove to us beyond a reasonable doubt that it doesn't?

Even if there were no proof of cancer, once there is evidence of harm, need we discount it? Cancer is only one of many diseases and disorders affecting Americans today. Science, and much of society for that matter, is all too ready to ignore things that it fails to understand. For years, complaints about fluorescent lighting in office spaces and VDTs were thought to be spurious and self-indulgent. People complained of headaches, fatigue, eyestrain and depression. But, for the most part, these are things that science cannot measure. Does that mean that they do not exist and that we should not do anything about them until science is capable of explaining them to us?

Ryan offered some suggestions on how a person can protect himself from electromagnetic radiation. The list that follows is simple but effective.

•Change over to traditional tungsten light bulbs from fluorescent fixtures or any bulb types that require ballast devices (i.e., metal halide, sodium vapor, etc.).

•Avoid using electric blankets completely.

•If you must use microwave ovens, have them rechecked in your home, after their normal factory inspection.

•When watching TV or using a computer video–display terminal, try to sit as far away as possible. Take frequent breaks to avoid eyestrain and back and neck tension. Also, you should use a computer radiation shield (a clear lead/acrylic screen that goes between you and the video screen) to block video–display radiation.

•Wear a Teslar watch, which collects destructive waves of extremely low frequencies, converts them to an eight Hz field (the pulse of theeEarth's natural field), and surrounds you with the eight–Hz cocoon to protect you from other electromagnetic fields.

Chapter 11
The Toxic Time Bomb—How America
Deals with Its Garbage

Up in the thick forests of the Pocono Mountains, life moves against an idyllic backdrop. Dotted with resorts, the area is a permanent home for a handful of people who have exchanged the conveniences of urban dwelling for a lifestyle that they hope is healthier in all ways.

In one small village, Pocono Summit, the residents were alarmed to find that a small clearing in the woods near their homes harbored a deadly secret. Wearing gas masks and protective clothing, workers from Pennsylvania's Toxic Waste Investigation and Prosecution Unit excavated fragments of twenty 55-gallon steel drums from the clearing. Corroded and empty, the drums had once contained a stew of toxic chemicals and industrial waste that was buried secretly and illegally in the woods. Officials feared that 300 or more barrels might have actually been buried. Water samples taken from 10 feet below the surface of the clearing revealed concentrations of toxic chemicals, up to 1,000 times higher than limits considered safe. Many of them were suspected carcinogens, and they had been seeping into the ground for at least six years.

Most of the residents of Pocono Summit draw their drinking water from wells adjacent to the contaminated site. One resident who lives near the clearing said that his well water got so brown that he would not even use it to wash his car. A neighbor said that his water seemed pure enough, but wanted to have it tested. His wife was even more vehement about the toxic-waste dumping, which was eventually linked to organized crime. "This is our home for the last thirty-three years," she said. "I can't believe somebody could do that and not know the danger."

Their outrage and frustration is understandable, but how worked up should we get over the problems that occurred a few years back in a tiny mountain village? After all, the problem is not in *our* backyard. There is now legislation to protect us from things like that. Or, if it does occur, there is plenty of money in the Superfund, and the EPA will make sure that everything is cleaned up and we are all compensated. Right? Well, unfortunately, it doesn't quite work like that.

Understanding the Scope of the Problem

America generates more garbage than any other nation on Earth. This is meant quite literally, although when it comes time to look at solutions to this monumental mess, we may discover that it is true symbolically as well. In other words, the problem may be as much about the nature of what we are producing as it is about the quantity of it that ends up in our trash heaps. Individually, we generate about 1,300 pounds of trash per person each year. American industry is responsible for nearly 250 million tons of toxic, corrosive, or ignitable waste. And as with so many other environmental issues, waste disposal in America has grown to crisis proportions in the 1980s. Some 80 percent of our solid waste is currently being dumped in landfills, but sites are rapidly filling up and are simultaneously leaking toxic chemicals into the nation's drinking water. Incinerators are costly, and fears of toxic emissions and ash make burning anything but the ideal solution. As we saw with the garbage barge from Islip, Long Island, that traveled more than 6,000 miles in search of a foreign dump for its putrid contents, only to return home 187 days later still holding its 3,186 tons of trash, we cannot look to other countries to passively accept America's trash. And while there are suggestions that we may be able to use space technology to catapult our waste outside the earth's atmosphere, most reasonable people are rejecting this as being at least as foolhardy and shortsighted as our landfill and incinerator solutions have proven to be.

What is perhaps the most startling about the waste dilemma in America is the conspicuous lack of leadership coming from government. One concerned citizen, in a letter to *The New York Times*, wrote that when she lived in Italy in the early 1980s, the public was frequently notified about procedure for local trash removal. The sanitation problem was kept under control. Since she moved back to New York, she had seen no disposal and recycling guidance provided by the city. She pointed out that 8 million people could make a difference in the cleanup of New York City and concluded, "It is unacceptable to inform us of pollution doomsday without giving us some city-wide plan to reverse the trend."

Although that woman wrote about New York City, what she says applies equally to most localities as well as to the federal government. We are simply not getting the kind of leadership that we need in order to deal effectively with a problem of the magnitude of our current trash dilemma. Furthermore, there is also evidence that government can be among the worst polluters.

Additionally, the problem of government-generated pollution goes way beyond instances of illegal corner-cutting—the magnitude of the nuclear-waste problem from weapons testing itself is staggering. In July 1988, congressional researchers revealed that the Department of Energy's prediction that the maximum cost of nuclear waste cleanup was far too low. While the DOE had set $110 billion as the ceiling figure, the General Accounting Office (GAO) said that the cost of restoring the environment, disposing of the waste, and building plants to continue bomb production would cost an estimated $175 billion and that even that figure could be low. According to the GAO, even with a $175 billion budget, which would cost American taxpayers more than $2,000 per household, the 17 major sites and the number of smaller ones would still be too dangerously contaminated to permit reuse.

The magnitude of this problem provides us with a perspective that is often overlooked—the costs of cleaning up after ourselves. These costs, for instance, are never cited when the Department of Defense draws up its proposals for new weapons.

Nor is reference to costs such as these made in discussing the price tag attached to the manufacture of toxic chemicals, or the disposal of raw sewage into the ocean, or the indiscriminate dumping of waste into landfills. Although in the past we could avoid looking at these costs in the shortterm, it is now becoming increasingly apparent that we cannot continue to do so. For instance, do we even know what the cost of dumping medical waste directly into our sewer systems can be? We may not, but can we afford to ignore it? Consultants for the Natural Resources Defense Fund estimate that 1.7 million pounds of infectious waste are generated each week in New York hospitals. Liquid portions of this waste such as blood from kidney dialysis, chemotherapy, and infected body fluids from AIDS, cancer, and hepatitis patients, can legally be dumped into the sewer systems of the state. Much of this waste is funneled out into the ocean off the New Jersey shore. Prompted by high bacteria counts and medical-waste washups in the summer of 1988, the state launched a campaign costing close to $3 million designed to quell fears that were scaring people away from New Jersey's beaches. The 1988 scare cost New Jersey an estimated $800 million in lost tourist revenue. These costs, of course, only include a partial loss of revenue for one state in one year. They do not begin to cover the long-term loss of revenue to individuals for decreases in property values and costs of medical care for people who become ill from the contamination.

The ubiquitous plastic container is also causing major disposal problems. Plastic can be found everywhere in the packaging of American consumer goods—from laundry detergent jugs to fast-food containers to yokes that hold beverage cans together. And as if there were not already enough of it *in* our garbage bags, most Americans use plastic to contain their garbage as well. Apart from the sheer volume that plastics add to our trash heaps, they pose additional problems. They do not biodegrade in our landfills, and if incinerated, they give off toxic fumes. While there has been some government action to lead the campaign against plastics, the action has been limited. New Jersey, for instance, is proposing a three-cent tax on nonbiodegradable plastic containers. But the $80 million that would be generated by this tax is not even earmarked for environmental purposes, making this look more like a fiscal trick than an honest effort to combat the ever-growing mounds of plastic waste. It contrasts, for instance, a proposed bill in St. Paul and Minneapolis that would ban most throwaway plastics. Similar legislation banning the use of polystyrene plastic to package food has already been adopted by local officials in Suffolk County, Long Island. Particularly remarkable about legislation such as this is the wide public support it receives. Government and industry spokespeople repeatedly insist that the American public just does not want to be bothered with changing, but this is not the case. Minneapolis council members stated that they were flooded with phone calls supporting the passage of the legislation banning the plastics. In fact, much of the support came following a campaign by plastics manufacturers and supermarket retailers opposing the measure and urging people to speak out against it. The ad campaign has just the opposite effect. According to one Minneapolis City Council member, the public support for the measure was amazing. He had more calls on this one issue than anything else that has ever been before the council.

Unlike other countries, America is still dragging its feet when it comes to handling and recycling plastic materials. Supermarkets in Austria use plastic bags that are biodegradable within 90 days. Trash in Rome and Paris is bagged in paper, not plastic. The EPA estimates that at our current rate of usage, plastics will form 15 percent of the country's solid waste by the turn of the century.

America is facing a problem of epic proportions with its ever growing mountains of garbage. Not only are landfills simply not capable of holding any more trash, but the more we learn about garbage, the more apparent it becomes that much of it cannot just be thrown away. Bethlehem, Connecticut, for example, is facing a problem that is just beginning to plague the nation as a whole—a dump yard filled with rusting used appliances that haulers and scrap-metal dealers are now refusing to take off their hands for fear that the equipment is riddled with PCBs. These chemicals, used as insulating materials in appliances like refrigerators and microwaves before they were banned in 1979, are believed to be potent carcinogens. When they are shredded for scrap, these old appliances release unsafe levels of PCBs, which can cause liver damage and skin lesions. Today, EPA requirements state that any material with more than a 50 parts per million concentration of PCBs must be disposed of as *hazardous* waste. Cities are slowly managing to cope with the crisis. Public workers and appliance dealers are being trained on how to safely remove the capacitators containing the PCBs so that the appliances can be safely scrapped. But nevertheless this situation is teaching us that some things just cannot be "thrown out with the trash." As our knowledge of the hundreds of other chemicals we have used over the past 40 years grows, it is probable that this situation will become more and more prevalent.

Landfills—No Longer "Bottomless Pits"
About 80 percent of all of America's solid waste ends up in a landfill. This trend is on the wane, but that is primarily because the dump sites are becoming filled. According to the Natural Resources Defense Fund estimates, the number of landfills since 1984 has decreased by 30 percent.

As recently as 1983 these areas received 98 percent of the nation's toxic waste. According to the GAO, every county of every state contains some type of toxic-waste dump—more than 378,000 of these sites across the country may pose a serious threat to public health.

A variety of studies, hearings, and independent research has brought to light the fact that the most common method of toxic-waste disposal is also the least safe. Experts agree that even the most advanced landfills in use today will eventually leak, and hazardous substances will seep into the surface environment or underground waterways.

Although landfills are designed to be closely watched for decades, toxic waste can remain hazardous for centuries. Environmental-medicine specialists contend that the dumps are just not secure. They further note that hazardous-waste landfills are impractical and unsafe unless you're prepared to spend huge amounts of money to line and hermetically seal underground caskets of waste. The linings themselves must resist degradation for the hundreds of years that the chemicals last. A survey conducted by the EPA in 1983 revealed that many storage and treatment facilities currently in use have no linings at all.

Landfills are popular because compared with the cost of incineration and other alternative disposal methods, dumping toxic wastes in the ground is a cheap solution to the immediate problem. If toxic dumpers were made fully liable for cleaning up leaks and for financing medical care for persons harmed by the substances, the use of landfill would lose its attraction. If we were to tax hazardous wastes to reflect external costs and add the right mix of ecological incentives and regulation, we would encourage a move to other alternatives and to less use of toxic chemicals. Groups such as the Chemical Manufacturers Association, whose member companies generate 71 percent of all hazardous waste, argue that landfilling can be done in an environmentally sound manner, but even that interest group admits that we are headed toward a condition where landfills will be used as a last resort.

According to the EPA, less than one third of the nation's toxic waste dumps meet requirements for monitoring underground water supplies near their sites under current toxic-waste disposal laws. Gene Lucero, head of the EPA's toxic waste enforcement program, believes that many dump operators would decide to close rather than absorb the expenses of making landfills safer, installing adequate monitoring systems, or getting sufficient liability insurance. Congress passed amendments to the Resource Conservation and Recovery Act in 1984 that required all hazardous-waste facilities to comply with financial liability standards or close. The virtual disappearance of competitively priced environment-impairment liability insurance has already forced many dumps that actually comply with federal standards to shut down. This scarcity of pollution liability insurance can be traced to shrinkage of the insurance industry's reserves, which has increased premium rates and reduced new policy writing for specialized areas such as environment impairment. Today, two thirds of our toxic waste dumps still do not meet their financial responsibilities as outlined by the EPA.

Congress intends to close all the dumps that fail to meet requirements, but so far it has been slow to act. "[The] EPA and the Justice Department do not have the resources to prosecute people who violate the law," said EPA waste management program official Hugh Kaufman. "We do not have the money or the people." A poll taken for *Time* magazine revealed that 64 percent of Americans would be willing to pay higher state and local taxes to fund cleanup programs in their areas. Seventy-nine percent felt that not enough was being done to clean up hazardous waste.

Like some noxious bogeyman, the specter of toxic waste seems to have leapt into public consciousness suddenly even though unintelligent disposal of hazardous waste has gone on for centuries. By the time the government got around to surveying the situation, the amount of toxic waste actually being generated was found to be four times the amount estimated. The OTA reported that the 10,000 priority toxic waste sites across America could easily cost $100 billion to adequately improve. The Reagan administration proposed only $5.3 million for a five-year cleanup program, and opposed a new tax on manufacturers proposed by Congress, that would have provided funds for cleanup.

The EPA has 850 toxic-waste dumps on its national priority list of sites that pose a potentially irreversible threat to public health, far fewer than the OTA's 10,000. In 1980, the agency declared the toxic-waste situation a "ticking time bomb primed to go off." That year Congress created the Superfund, a $1.6 billion five-year program intended to clean up thousands of leaking dumps endangering groundwater. During its mandate only six sites were cleaned up, and with questionable success. At two of the sites, Greenville, Mississippi, and Cleveland, Ohio, cleanup consisted of hauling nonleaking drums of chemicals to nearby landfills. In 1982, when Pennsylvania's Susquehanna River was polluted by the illegal dumping of toxic-waste water by a small company, the EPA supervised the cleanup and removed the site from its priority list. One-hundred-thousand gallons of chemical waste were stirred back up to the surface of the supposedly clean river, however, when Hurricane Gloria pelted the area with torrential rain. The EPA hauled off 1,500 leaking drums and a foot of topsoil from a former chemical dump in Baltimore. Contamination of the area, however, was found as far as 15 feet below the surface, and no further action has been taken to keep the chemical from infiltrating the groundwater or a nearby river. At the fifth site, 68,204 cubic yards of contaminated topsoil removed from a golf course that bordered the Velsicol Chemical Company dump near St. Louis, Michigan. The EPA merely hauled the tainted dirt to the Velsicol plant yard, and considered the job completed. The National Campaign Against Toxic Hazards claims that fewer than 10 percent of the sites on the priority list have got any attention at all from the EPA. According to the group, millions of Americans will wait decades for the EPA to clean up their poisoned communities.

Superfund expired in October 1985. It was widely recognized at the time that its five-year existence had been a billion-dollar boondoggle. Furthermore, the EPA, which supervised the cleanup, was ·found to be right in the middle of a scandal involving coverups, perjury to Congress, political patronage, and character assassination.

The Superfund fiasco started to come to light when EPA official Hugh Kaufman, special assistant in the Hazardous Waste Division of the agency, gave testimony before Congress that the EPA was failing to fulfill its obligations under the Superfund law. According to Kaufman, in retaliation for his revelations concerning the EPA, the Reagan administration proceeded to discredit him, using such tactics as investigating his sex life and giving him poor performance ratings at work. Kaufman's allegations sparked a congressional investigation that ended up confirming not only his claims of harassment, but also revealed gross mismanagement by the EPA in its handling of Superfund monies and programs.

The Reagan administration's harassment tactics ended up backfiring, however, and only prodded Congress to investigate more diligently what was happening in the upper echelons of the federal government. The fiasco was under way. First, the administration fired EPA assistant administrator Rita Lavell, who was in charge of starting up the Superfund project—some believe as an attempt to distract attention from what was really going on in the EPA. She was also convicted of perjury for denying her involvement in the EPA's dealings with a notorious waste dump in California. Then, during the course of the investigation, the administration attempted to enact an executive privilege on all documentation requisitioned by Congress. For her attempts to withhold vital information and impede the congressional investigation, the head of the EPA, Anne Burford, was found in contempt, giving her the dubious distinction of being the highest-level government appointee ever to be found in contempt of Congress. She resigned as head of the EPA in 1983.

As a result of the documentation that Congress finally received, it found that not only was Superfund not working, but the program had not been used to protect public health, the purpose for which it had been enacted.

Congress also found that one third of the Superfund money supposedly spent on cleanup could not be accounted for. It further discovered that nonpublic, closed-door negotiations were taking place between major polluters and top EPA officials. At these meetings, sweetheart deals were transacted where a minimal amount of money was paid to the government for a minimal amount of cleanup in exchange for a letter from the EPA relieving the polluters from any future liability for damage. In this regard, Lavell's daily calendar said it all—she was being wined and dined for lunch and dinner daily by the chief lobbyists for the Fortune 500 chemical companies.

During all this, the EPA also suffered a 23 percent cut in its budget and lost 19 percent of its employees. One thing that did come out was a sharpening of the public's attention to the magnitude of the cleanup job in store for America—a job that could not simply be ignored. Consequently, the Superfund was reauthorized in 1986, with the mandate to permanently eliminate toxic hazards from the sites on the EPA's list. Unfortunately, however, in a report issued by the Natural Resources Defense Council that examines the EPA's performance during 1987, the agency receives less than commendable marks. Testifying on the report before the House Subcommittee on Oversight and Investigations, attorney and coauthor of the report Doug Wolf said: "The American public had high expectations for the Superfund program once the amendments of 1986 (SARA) became law. This is because SARA dramatically changed the emphasis of the Superfund program from containment to permanent treatment, and gave the EPA almost 9 billion dollars to carry out this new mandate. Unfortunately, we are here today to report that, based on our comprehensive review of EPA's 1987 Superfund cleanup decisions, these high expectations have not been met."

Many people don't know about the presence of dumps in their neighborhoods and never make a connection between ill health and ill-managed toxic waste. A study conducted by the University of Medicine and Dentistry of New Jersey documented a startling correlation between toxic dumps and the cancer death rate in surrounding areas. Examining the mortality rates of 200 municipalities, researchers pinpointed a concentration of cancer deaths in a northeastern corridor of the state. In the 20 cities and townships within the area, the death rate for esophagus, stomach, lung, larynx, and colon cancer exceed the national average by *50 percent.* According to the researchers, these areas are characterized as urbanized by being highly industrialized and having a high population density. The heavy concentration of toxic-waste disposal sites virtually overlaps the 20 areas of excessive mortality.

The study contends that the disposal of hazardous wastes is the major problem facing this country over the next 20 years, and that the public is neither as concerned nor as constructively aggressive as it should be with the matter. According to Dr. Donald Louria, chairman of the university's department of preventive medicine and community health, "If a cancer- or other disease-producing substance leaks out of an industrial site or waste dump and gets into the water or food supply, the cancer may not appear until ten years later."

The Mississippi River has been described by water experts as America's longest sewer system. For years, chemical and other wastes have been deposited in the river's delta wetlands; some of the local parishes in the region have rates of colon cancer that are among the highest in the nation. Thousands of pounds of toxic chemicals are dumped each month into the Hudson River, New York's most important waterway and a source of drinking water for state residents. In most cases, the dumping is illegal, but some of it is actually sanctioned by the State Department of Environmental Conservation. "If in fact we set zero-pollution standards, we would not have any business left in the country," said Daniel Barolo, director of the department's division of water. The state routinely issues permits that allow dozens of industries to dump toxic waste, including benzene, toluene, cyanide, and lead in the Hudson.

Although the EPA concedes that the problem is much bigger than they ever suspected, the agency contends that they are taking the best course. The EPA feels that we can't wait around until we have the ultimate answer, because garbage is still being generated today, and we have to deal with it today. So yes, if we don't want to store waste in drums all over the country we are going to put it in landfills that may leak someday.

Alternatives

One option that the EPA seems slow in coming to grips with is to push for tougher regulations and restrictions on the generation of toxic materials in the first place. While there is an undeniably great need to clean up existing sites, unless the billions of dollars we are spending in the process are simply seen as a stopgap measure, no permanent cleanup can be expected without a limit to the amount of new toxins that need to be landfilled. Even when the EPA sets new strategies to handle the nation's garbage crisis, these approaches are far from hard-hitting. For example, in 1988 we dumped 80 percent of our garbage into landfills, 10 percent was incinerated, and 10 percent was burned. The agency's new broad strategy to deal with the nation's garbage crisis is to decrease the nation's solid waste by 25 percent. While the announcement of this plan was seen as a step in the right direction, it was made just weeks before the 1988 presidential elections. Moreover, while the proposals sounded good, the EPA announcement lacked any regulatory clout, including budgetary allowances that would take the proposals from paper to reality.

Very simple and inexpensive programs, which have already been implemented by a number of municipalities, have decreased the amount of garbage by more than the 25 percent.

Incinerators

Unheard of prior to 1970, incinerators that transform our mountains of garbage into mounds of ash are now common in the United States and also called "resource recovery" systems. There are currently about 100 incinerators in operation, and plans are on the table for the construction of approximately 75 more. On the East Coast, construction proposals for these garbage-burning incinerators are particularly numerous. Among these are massive plans for 18 incinerators in New Jersey, 14 for New York State (8 for New York City alone), 8 for Connecticut, and 14 for Pennsylvania.

The construction of these gigantic ovens is being pushed with a considerable amount of zeal by industry representatives and politicians alike. They are presented to the public as the *magic bullet*, the solution to all of our waste-disposal problems. And a painless solution at that. With incinerators, we do not have to alter our consumption or waste-disposal patterns at all. We can continue to throw away as much as we want, and trash can still be discarded and hauled away in the same manner. What is not discussed, however, are the financial, environmental, and health costs associated with these furnaces.

In terms of construction costs alone, incinerators are exorbitantly expensive. An average-size incinerator, one of about a 1,000-ton capacity, which could accommodate a medium-sized city, costs around $100 million. New York City's Department of Sanitation estimates that building the eight incinerators needed to handle about 70 percent of its garbage—some 18,000 tons of trash per day—will require an initial capital investment of $3 billion. One company involved in the construction of these garbage incinerators estimates that there is a $20 billion market for the construction and operation of these facilities over the next 15 to 20 years. Nor are initial outlays the only costs associated with incineration; the maintenance and operations can be very high. In Dade County, Florida, these costs for a 3,000-ton facility cost $6.75 million annually. In Ames, Iowa, the maintenance and operation amounted to $807,000 for a 180-ton facility. But it is not just the exorbitant financial costs that have brought many people to view the claims about incineration with a healthy degree of skepticism.

According to environmental activist Madeline Hoffman of the consumer group Stop Incinerator Now, all garbage incinerators currently in existence, both in the United States and abroad, have been shown to emit furans and dioxin, a component of Agent Orange that we now know to be an extremely potent carcinogen. Citizen fears about these chemicals has already caused a 2,000-ton, $130-million incinerator to close in Hempstead, Long Island, despite the fact that state laws mandate the closure of all landfills by 1990. Garbage incinerators also emit gases like hydrochloric acid, nitrous oxide, and sulfur oxides, which not only pose a threat to human health but also are the primary cause of acid rain. Nitrous oxides are of particular concern because they are also believed to play a significant role in the depletion of the earth's ozone layer. We also know that incinerators will emit heavy metals such as lead, arsenic, and cadmium, many of which have been linked to neurological disorders and cancer.

Of major concern to environmentalists is that the East Coast construction proposals are being considered on a massive scale without any prior studies to determine the cumulative impact on the health and safety of residents. Moreover, all of this is taking place in one of the most populous and heavily industrialized areas in the country, where pollution is already a serious problem. Says Ms. Hoffman, "Before we embark on spending millions and millions of dollars on the construction of all these incinerators, it is important that we know what kind of pollution we are already exposed to, and what the cumulative impact of these incinerators will be.

"Incineration and landfills often go hand-in-hand and reinforce the current system of garbage collection and disposal. Both the people who operate landfills and those who use them as dumping grounds for their materials feel threatened by the changeover, or the pressure to close down the landfills. They are looking for some way to continue their business as usual. An incinerator allows them to take the same materials, hauled in the same way, bring it to an incinerator to be burned. This also allows for the continued life of these landfills because incineration of garbage generates a toxic ash. That ash must be landfilled. So, where as before there was pressure to close the landfills, now there is pressure to keep them open. Furthermore, the waste haulers who are benefiting under the current system of garbage collection and disposal will be able to continue to do that."

Ken Bruno of Greenpeace comments on the problems associated with the toxic ash that results from garbage incineration: "The ash from garbage incinerators is very toxic. Philadelphia provides a good example of the absurd proportions that toxic incinerator ash can reach. That city had tremendous problems trying to dispose of its ash because of its toxicity. Landfills in Ohio, New Jersey, and Pennsylvania refused the ash from Philadelphia, so it started looking overseas. Just like the infamous garbage barge from Islip, it sent barges around the Caribbean and all the way to Africa looking for a home. In one case, a tanker traveled for over two years to the Cayman Islands, to the Bahamas, to Columbia, to Panama, looking for a home. It ended up dumping about three thousand tons of it on a beach in Haiti under the guise that it was fertilizer imported by two businessmen. People in Haiti had no idea what this stuff was. They don't have incinerators and are not used to handling it."

Many people think that the solution to the pollution problem in incinerators is to put air-pollution devices on them. Unfortunately, the better your air-pollution controls, the more toxic your ash ends up because this stuff has to go somewhere—it just doesn't disappear. That goes for the heavy metals as well as the dioxin.

Until Americans realize that there are possibilities for garbage disposal that neither require enormous capital expenditures and a virtually irreversible contamination of our environment, the debates surrounding incineration and landfills will continue to rage at all levels of government. Most politicians are simply not willing to take the leadership on this issue. In New York State, for example, after two months of debate on solid-waste legislation, the state legislature ended up with compromise legislation that encompasses virtually every aspect of garbage disposal, including transportation of waste and training of incinerator operators. Key provisions of this legislation include:

•All landfills must have double liners with drainage systems at each level—purportedly to prevent contamination of groundwater. The landfill site at Fresh Kills, Staten Island, which has *no* liner, is exempted from the regulations.

•City proposals for the construction of new incinerators must be accompanied by recycling plans.

•Limits are proposed on emissions like dioxin, particulates, and acid gases.

•Ash from *new* incinerators must be disposed of in landfills with at least one liner. Ash from existing incinerators can be disposed of in landfills having no liner at all!

But even in a state with one of the most dire waste-disposal problems in the nation, these regulations are anything but comprehensive. While there is some requirement for cities to come up with recycling plans in order to receive permits to build new incinerators, the state regulations are conspicuously lacking any guidelines for sorely needed statewide mandatory recycling laws. What is equally alarming is that after months of negotiation in the face of evidence on the toxicity of incinerator ash, state officials elected not to treat the ash as hazardous material.

Fortunately, however, there are some public figures who are showing leadership and common sense when it comes to the issues of waste disposal in general, and incinerators in particular. Suffolk County official Mike LaRossa is one such figure, and he provides a good example of what a difference can be made when politicians are willing to work with, rather than against us. "There was a group of people that had been mandated by the state to advise us as county officials as to the pros and cons on incineration. Their original position had been in favor of the incinerators. We gave them a free hand to take another look at the situation—one year with unlimited funds to look at all the alternatives. In the end, that very same group that had voted unanimously to introduce incineration for our area reversed their decision and voted unanimously against it. We have now become the first county in the state to repudiate this approach. We are having a significant impact on our Department of Environment Protection because other counties around the state have been contacting us, both to gain from us the experience that we have had in how to educate the public and get the public involved, number one, and secondly, what were the results of our tests so that they can use our documentation to support their desire to turn against incineration."

Ocean Dumping

Thus far, we have been discussing mostly the problem of solid-waste disposal. But there is another kind of waste-disposal problem that is causing just as much environmental damage—the waste that goes down our drains and our toilets and into our sewage systems. It is this type of waste, often dumped directly into waterways without any prior treatment, that has irretrievably contaminated many of our rivers and is now threatening our oceans as well.

In southern New Jersey 9 million tons of contaminated toxic sludge is taken each year and dumped into the ocean. This has been going on since 1924. Up until last year, it was dumped in 80 feet of water 12 miles off the coast of Seabright, New Jersey.

The situation finally got so bad that even the EPA realized that it had to do something about it. Its solution was to move the dump site out to 106 miles. What they figured was this: where the Continental Shelf starts 100 miles out, the water goes from 600 feet to 6,000 feet deep. There is a cliff out there. They figured that all they had to do was to take all this sludge out there and dump it over the cliff. But they did not do any type of environmental-impact statement in that area. If they had, they would have found that it is a very complex area. The Gulf Stream swings into the site where the stuff is dumped; the shore currents carry fish eggs out into that area; the contaminants are spun out into a 45,000-square-mile area from Cape Hatteras up to Cape Cod. Moreover, most of the sludge does not go down 6,000 feet, because different layers of water have different densities. In the summer the first level out at site 106 is usually about 75 to 150 feet. Imagine dropping down through 75 feet of water and hitting a layer of vegetable oil. It will not stop a cannonball from going to the bottom, but it will stop just about everything that is dumped at site 106, all of this contaminated sludge. What that means is that 25,000 tons a day of sludge is being dumped into an area where the real floor is only 75 to 100 feet down. All of the dolphins, sea turtles, all of the fish eggs of 200 species are washing around in there and being contaminated. In tests performed by the Oceanic Society and Greenpeace, one year's total equals 900,000 pounds of copper, 5,000 pounds of mercury,

9,000 pounds of cyanide, 8,000 pounds of arsenic, 117,000 pounds of nickel, 1.5 million pounds of zinc, and 18 million pounds of iron. All of this is going into a fairly small patch of ocean out there and is being spun out over 45,000 square miles.

Bottlenose dolphins are dead, shellfish, and lobsters have burn holes and lesions, other fish have fin-rot disease, and people are getting sick.

Recycling and Composting

Contrary to what most American politicians are saying, our waste-disposal problems are not insurmountable, nor is the American public apathetic or unwilling to participate in programs designed to decrease our trash crisis. In fact, for the most part it is citizens, not public officials, who have taken the matter to task. In San Jose, California, for instance, about 60 percent of households participate in a voluntary recycling program. Says Richard Gertman, San Jose's recycling manager, "It would be cheaper for us to throw it all out as garbage, but people love to recycle. They can't control air pollution or traffic. But through this program, they feel like they can do something." We do not agree with certain of Gertman's statements—recycling is not more expensive if you take into account the environmental and health costs of disposing of waste in conventional manners; nor is it less expensive for cities that, unlike San Jose, have already used up all existing landfill space. It is also not true that people have no control over air pollution and traffic. But Gertman does make the important point that the American public is not opposed to taking an active part in cleaning up its environment. This was also seen in Seattle, where almost 60 percent of households were participating in a recycling program within six months of its inception. Today, nearly 28 percent of Seattle's garbage is recycled, and the city hopes to double this amount in the coming years.

The reality of trash disposal is quite different from what government officials, industry spokesmen, and even the press are telling us that it is. For example, one day preceding an article that discussed the success of recycling efforts, *The New York Times* published a commentary that concluded that "recycling will never replace landfills as a solution to the trash problem. Experts say a recycling goal of twenty-five percent of American trash is optimistic." Not only are statements such as these directly refuted by current reality— it is not even wishful thinking to recycle up to 28 percent of household garbage in an existing and voluntary program—they also do nothing to encourage further participation. Industry is particularly adept in issuing doomsday alarmist statements. Concerning Minneapolis and St. Paul ordinances to ban the use of plastic food packaging, Joel Hoiland, president of the Minnesota Grocers Association, says that a local food chain estimated that would cost $1.5 million per store to supply alternatives to plastic containers. Roger Bernstein, director of state government affairs for the Society of the Plastics Industry in Washington, described the ordinance as "insane," and a financial hardship for grocers and fast-food retailers. "I would have to harken back to Prohibition to think of a precedent. We don't even ban unsafe products in this country," he said. But on the other end of the spectrum, McDonald's, long a target of environmentalists because of its use of polystyrene hamburger boxes, recently instituted a recycling program capable of eventually reprocessing 65 million pounds of the plastic containers.

Tom Webster, a research associate at City University of New York, conducted some very promising experiments in recycling and composting in the Long Island community of East Hampton. His research shows that up to 84 percent of all solid waste is capable of being disposed of using current technology that is safe, nonpolluting, and does not require incineration. In East Hampton, New York, he enlisted a group of 100 volunteer families who operated a source-separation program for 10 weeks. They were able to accomplish an 84 percent recycling rate—meaning that of all the trash that those families generated, they were able to recycle or compost 84 percent.

By source separation we mean separating materials up front. Put the food wastes and things like yard wastes in one container; the recyclables like bottles, cans, paper, and sometimes plastics in another container; and then everything else in a third container. It is important to be careful to keep toxic materials at a minimum, first by not buying them in the first place, and when you absolutely need them, by keeping them separate in this third container. By doing that, you can take the food and yard waste and compost, making an earthlike material that is great for gardens. The recyclables can then be separated at a central facility and turned back into paper, glass, and cans. The third container accounts for about 15 percent, and we want to look at its composition and see what changes we can make so that ultimately it all can be either compostable or recyclable. With these techniques, it is feasible to recycle at least 70 percent of the garbage and potentially as much as 90 percent or more 10 years down the road.

The technology is in existence and operating today to sort and separate the metals, plastics, paper, and so on. That technology costs far less than incineration, number one, and it is totally effective, number two, and the repair, upkeep, and operation of it is far less than incineration.

Recycling and composting techniques have also been tried out in a number of American cities as an alternative to dumping raw sewage into waterways. In addition to an exemplary recycling effort in Seattle, the 100,000 tons of sewage produced each year there is processed and used as fertilizer, which stimulates trees to grow much faster than normal. Like New York and New Jersey, Seattle used to dump all of its sewage directly into its surrounding waters. Now, the Puget Sound is cleaner because the city recycles 100 percent of its sludge, most of it for the fertilizer, which is in high demand by private timber companies. Seattle is not alone: Los Angeles mixes its sludge with sawdust and sells it as compost, and Milwaukee sells its sludge as a lawn fertilizer named Milorganite. There is, however, a caveat here—without regulation of what substances can be thrown down the drain or adequate treatment processes, these recycled products should be used for limited purposes. For instance, if toxic chemicals are dumped down the drain, fertilizers like Milorganite may contain residues of these chemicals and could cause health problems for children playing on lawns on which the fertilizer was used. The same holds true if unregulated hospital wastes are discarded down the drain.

Arcata, California, has initiated one of the most effective waste-treatment systems in the country. In this coastal northern California town of 15,000, engineers have constructed some 154 acres of marshes, lagoons, and ponds that cleanse the city's waste waters. At the same time, the wetlands provide a habitat for thousands of sea birds, otters, and other marine animals. Based upon a solid understanding of marine biology, this system is considered by many to be the most innovative and effective in the nation, while at the same time costing practically nothing. The system is beautifully simple and in synchronization with the ecostructure of the region. First the water passes through oxidation ponds where the effluent is broken down into its component parts of hydrogen, oxygen, and carbon; then it moves on to be filtered and cleansed in marshes; and then, moves on to irrigate and nourish other wetlands. The waste water refuge, which now draws thousands of birdwatchers each year, has no smell other than that of salt air.

Recycling and composting not only provide viable alternatives to waste disposal, they also make sense economically and environmentally. It is estimated, for example, that the recycling of aluminium cans to produce other products can save as much as 92 percent of the energy required to produce the same products from virgin bauxite ore. The National Association of Recycling found that in 1979 more than 300 million barrels of oil were saved by the recycling of paper, metal, and other products. Obviously then, with a concerted effort to recycle large amounts of garbage, we could not only be solving our garbage problems but also decreasing the gases released from the burning of fossil fuels that contribute to acid rain, the greenhouse effect, and air pollution.

Recycling also has other notable advantages. First, a recycling program can be started with almost no expense and can be well under way in a year or two, where a resource-recovery system can take many years to build and cost many millions of dollars. Another argument for recycling is that even if it does not totally replace incinerators, it can greatly decrease the volume of garbage that needs to be burned. Consequently, the incinerators can be built on a much smaller scale.

Waste Management—America's New Growth Industry

The collection and disposal of America's solid waste was a $20 billion industry in 1987. According to EPA predictions, under current trends, the industry profits have no where to go but up. The agency estimates that by the year 2000 Americans will generate an average of 45 pounds of garbage per person each week, almost double the 28 weekly pounds in 1988. This makes the trash industry one of the most potentially profitable in the country.

Twelve billion dollars, or over half of the total dollars spent in 1987 on waste disposal, was for collection and transportation. By any standard, this is not an insignificant amount of money, particularly for a necessity. There is little wonder then that there is so little incentive for industry and government to really solve our waste problems, since any real solution would entail a decrease in *volume*, and it is precisely this volume that fuels this $12-billion hauling industry. Among the top firms in the garbage business are the Chicago-based Waste Management, which took in more than $2 billion in revenues in 1987, and Houston-based Browning-Ferris, which made over $1.5 billion in revenues.

Additionally, the price tags attached to waste disposal have risen dramatically over the past years. In the five years between 1982 and 1987, the cost of unloading trash at a landfill rose from $11 per ton to $20, while costs of incineration nearly tripled from $13 to $34. In a way, this can be seen as a good sign for the consumer. Maybe once the garbage-disposal rates really skyrocket, people will begin to demand alternatives.

What is important for all of us to realize is that there are alternatives. They do not need to cost billions of dollars, nor do they have to contaminate our environment and ruin our health. All that is required is that each of us become involved—most importantly in urging our politicians to adopt approaches like that taken by Mike LaRossa in Suffolk County, or Tom Webster in East Hampton, or those in Seattle, Arcata, and San Jose.

Things You Can Do to Get Involved

There are a number of things that you can do right now in order to start helping rather than contributing to our environmental problems. While the actions suggested here are geared toward the problem of waste disposal, you will see that many of them will also be useful toward solving many of the other problems that we currently face.

One of the first things you can do is to get involved in recycling. It is really a very simple process, and at least to start, you can get by with just three, possibly four, containers—depending on how well organized your community is. One container can be used for cans and bottles. Most communities have recycling facilities that can separate these once they are collected. Another can or box is used for paper materials—old newspapers, junk mail, computer printouts. People also need to push their companies, particularly offices such as law firms and banks that generate tremendous amounts of paper, to also get involved in recycling. Then, whether you have one or two containers left depends on whether your community is composting organic matter or not. If so, then you use the third container for organic things, and the fourth for all the rest. If not, then you lump everything that is not paper or recyclable bottles and glass into the third container.

In terms of what you are contributing just by doing these simple separation actions, consider the following:

•Recycling paper products can save 70,000 trees each day;

•We throw away enough glass every two weeks to fill up the 110-story twin towers at the World Trade Center;

•In three months, we throw away enough aluminium to rebuild our entire commercial air fleet;

•Every minute of every day we throw away 42,000 plastic bottles, or 60 million each day.

There are also a number of ways that you can reduce the toxicity of what goes down your drains and into your garbage cans. Remember, this stuff has to go somewhere, so the trick is not to use it in the first place if you can avoid it. Anything that goes down your drain may either come back to pollute your groundwater or may be turning your favorite beach into a toxic soup. Now available are lists of natural, nontoxic products that can be substituted for commercial products. Here are just few of the suggested substitutions:

Replace:	With:
Ammonia-based cleaners	White vinegar or cider vinegar mixed with salt and water for cleaning surfaces. For bathrooms, use baking soda and water
Furniture or floor polish	1 part lemon juice with 2 parts olive or vegetable oil
Drain cleaners	Plunger, or flush with boiling water, 1/4 cup baking soda, 1/4 cup vinegar
Disenfectants	1/2 cup borax in 1 gallon water
Rug Cleaners	Sprinkle cornstarch on, then vacuum

For more alternatives to commercial chemicals, write the Household Hazardous Waste Wheel, Box 70, Durham, NH 03824-0070.

You can also have a great influence on the amount of garbage by making conscious consumer decisions. You can ask for paper instead of plastic bags whenever you go to the supermarket. It is also surprising how much plastic we do use when we shop. You can without any great effort cut down on the number of plastic bags that you use at the supermarket—not *everything* has to go into its own separate bag. Those that you do take you can reuse, rather than tossing them out and putting foods into purchased Baggies. You can also reuse your paper shopping bags as well as any plastic bags to line your garbage cans rather than throwing them in with the rest of the trash. This not only decreases the bulk of your garbage, but also lowers the number of plastic garbage bags you add to our trash piles.

Another thing that everyone can do is to ask for paper cups and containers for take-out food and beverages. You can also bring your own coffee mug or food containers to cut down even further on what is one of the largest sources of American trash and litter. In the supermarkets you can buy products that are packed in paper or cardboard rather than plastic or polystyrene foam—eggs, milk, butter, and so on.

In terms of laundry detergents, many powdered products contain phosphate unless they are marked. Liquid laundry detergents do not, so you may want to either switch to one of these products, preferably one that indicates that the plastic jug containing it comes from recycled plastic, or looking for a powder that does not contain phosphates.

There are currently two brands of biodegradable plastic bags on the market, Ruffies and Good Sense. Also there are biodegradable baby diapers called TenderCares. They are still about $1 to $1.50 more expensive than the commercial brands, but it is worth it in terms of the environment. As their market grows, their price will go down accordingly.

Resources:

For those of you who are interested in starting programs in your area or who require information on how to deal with toxic chemicals in your environment, you may contact some of the following sources:

•In New Jersey, contact the Grassroots Environmental Organization, Box 2018, Bloomfield, NJ 07003, (201) 429-8965. Elsewhere in the United States, contact the Citizen's Clearinghouse for Hazardous Waste, Box 926, Arlington, VA 22216, (703) 276-7070. This last organization assists organizations all across the country in fighting the battle against hazardous waste.

People who believe that they may have been exposed to toxic pollutants and would like to find out what their legal rights might be can contact the Environmental Action Foundation in Washington, D.C. at (202) 745-4879.

For information concerning recycling programs in your area, first try your local city council or sanitation department. In the event that they give you the runaround, which is often the case, you can call the Environmental Defense Fund's hotline at 1-800-CALLEDF.

Solutions to solid wastes can only be found through reduction in size of the waystream and in toxicity of the waystream. Avoid buying large things that you can borrow or that are reusable. You must think ahead to avoid creating garbage. Using recently available technology such as plastic trash bags treated with a cornstarch solution that allows them to biodegrade in a few years reduces the burden.

Be aware of not throwing resources away, and be aware that what you do throw away is not toxic to the environment. Look to recycle everything you can. Be aware of purchasing recyclable materials as opposed to buying items that are packaged in a way that cannot be used again. Wax or plastic linings can make paper nonrecyclable. Compost all organic, nontoxic materials. Buying in bulk, rather than individual sizes, reduces the amount of packaging materials we have to dispose of. Buy rechargeable batteries.

For more information contact:

The Environmental Action Coalition
625 Broadway
New York, NY   10012
(212) 677-1601

Part Four
What We Can Do About It

Chapter 12
Safeguarding Your Health

Taking a survey of today's environmental problems can be overwhelming and depressing. Many people look around, hear about the problems, not just of the environment, but also concerning issues of health or politics, and their response is to shrug their shoulders. "Oh, well," they say, "everything causes pollution. Everything causes cancer. . . ." Or, if the problem is one of government, "What do you expect? That's just politics. All politicians are corrupt." These answers are understandable in a world inundated with media coverage of disasters and death. When the evening news covers AIDS or cancer, for instance, we are told that the diseases are fatal, and then to confirm this belief, we hear many stories of people who are suffering and dying. Although we do not hear about them, there are in fact people who are surviving the illnesses. A news story covering AIDS survivors and what they are doing that is keeping them alive would leave people with a substantially different impression.

When we receive only bad news, we feel dwarfed by the seeming magnitude of problems and are left with a feeling of impotence. What can we possibly do that will make any difference? There are, in fact, many things that each of us can do, both in terms of maximizing our health and well-being, and also with respect to cleaning up the environment.

Surviving in a Polluted World

Every year Americans are bombarded with a myriad of new substances that enter into and become part of our daily lives—paints, laundry detergents, gasoline, hair sprays, pesticides, furniture polishes, and plastics. In addition to products routinely used in our households and businesses, we are only now beginning to under-

241

stand some of the long-term effects of postwar technology. We now know, for instance, that even small doses of the radioactivity released from nuclear–power plants and weapons facilities, even in small doses, can cause cancer and other health effects. Pesticides are not only poisoning our food supply, but our groundwater and our oceans. We have consumed so much, with so little attention to the long-term environmental effects, that we now are experiencing tragedies like the destruction of the Amazon forests, the depletion of the earth's ozone layer, and the deterioration of our air, soil, and water.

That is the bad news. There is no use denying that we live in a very polluted world. While there is no way to totally avoid exposure to environmental contaminants, there is still plenty that we can do to minimize the effects pollution will have on our health.

Medicine today tends to view illness and disease in terms of their symptoms alone. If you have cancer, then the tumor or lesions are equated with the illness. Treatment consists of trying to eradicate that symptom. The same holds true with other conditions. If you have a headache, you are told to take aspirin or Tylenol. If you have arthritis, you are prescribed an antiinflammatory drug. But rarely does medicine endeavor to get at the root of the problem. In reality the tumor or any other symptom is only the last stage of an evolving process of ill health.

With this symptom approach to illness, most doctors today fail to see the body as a whole in which each part is interrelated and interdependent.

When people hear that this food, this chemical, or this additive can cause cancer to this or that organ, they become overwhelmed. We've heard many people say, "Well, since *everything* causes cancer, what good does it do to even try to change?" If, on the other hand, we begin to understand the way our bodies work and focus our attention of total health rather than on the avoidance of a specific illness, we can take back our power, start healing ourselves and prevent illness.

Because modern medicine has proven ineffective in dealing with many of today's major illnesses such as AIDS and cancer, some people are opting for alternative methods of healing. At the core of these therapies lies a more holistic or integrated view of the body and how it interacts with the environment. Focus is placed on strengthening the body's immune and cleansing systems, so that the body can regain its innate ability to combat disease. According to Dr. Stanislaw Burzynski of the Burzynski Cancer Research Center in Houston, cancer cells are produced every day within our bodies. When our immune system is functioning properly, it identifies and destroys these cells. When we are overly stressed or in illhealth, however, these factors can diminish the body's natural cancer–fighting abilities. Dr. Burzynski and other alternative cancer specialists believe that in order to *cure* cancer, these stress factors must be brought under control, while at the same time must be built up the body's natural defenses.

Also important is the body's ability to detoxify itself through its major cleansing organs—the liver and the kidneys. Dr. Max Gerson found in his research and treatment of cancer patients that almost without exception people suffering from the disease had significant liver and kidney impairment. Accordingly, one of the main objectives of the Gerson therapy is to build up these organs.

Strengthening the immune system and detoxifying the body is fundamental to the treatment and prevention of all diseases. We are constantly assaulted by a wide variety of toxins, each of which acts as a stressor on the immune system. Mental and emotional stress also causes toxins to be produced within the body that combine with the effects of external pollutants to drain the body's capacity to ward off disease. While we may not be able to live within a pollution-free environment, we can take steps to minimize the effects of environmental toxins.

A healthy immune system is dependent on a number of nutrients. Among the most important are the vitamins A, B–complex, C, and E, and the minerals zinc, selenium, and magnesium. These nutrients not only help maintain a good defensive system, but work within the body to detoxify pollutants. Many nutritionists today are realizing that almost all toxins cause their damage to the body through what are called "free radicals." These are unstable molecules, lacking an electron in their outer shell, that can cause chain reactions within the body. One of the primary roles of vitamin C, for instance, is to clinch these free radicals by donating an electron and thus deactivating them and putting a halt to their harmful process.

For people who are concerned about their health and protecting themselves from environmental toxins, we strongly recommend learning about important vitamins and minerals. A good place to start is your local health–food store or bookshop. There are a number of books now available that will discuss the role of different nutrients in boosting the immune response. If the health–food store has knowledgeable personnel, they should be able to inform you on the different vitamins and minerals you may need. Remember that you must follow your own sense of what is right for you. If you are not a "pill taker," then buying a large number of supplements obviously is not going to do you much good. Multiple supplements such as B-complex (as opposed to the individual members of the B-vitamin family) are usually adequate for most people. Unless you can purchase reputable brands of vitamins from your pharmacy, I advise that you obtain your vitamins from a health–food store. By reputable, I mean companies that do not use artificial colors, sugars, binders, and preferably avoid common allergens like wheat, corn, yeast, citrus, and dairy derivatives.

Vitamin and mineral needs are individual and may vary according to your particular health conditions, stress, and age, so you may also want to consult a good nutrition-oriented physician in order to get the combinations and dosages that best suit you. There are a number of tests now available that will help in diagnosing your needs. A hair analysis, for example, reveals the levels of heavy metals in your system. If these tests indicate high levels of these toxins, your doctor can recommend certain supplements that work to chelate or bind heavy metals and remove them from the body. If you are suffering from a host of vague symptoms like fatigue, headaches, or depression that have not been helped by traditional medical treatment, you may also want to be tested for environmental allergies by a physician practicing clinical ecology.

For people who lack access to a nutritionally oriented physician, or who feel that they do not have specific needs, yet want to supplement their diets to protect against environmental toxins, the following supplements are easy to take and available at most good health–food stores:

•Vitamin A is synthesized in the body from its precursor, Beta-Carotene. Unless you have problems with this synthesization process, Beta-Carotene supplements are better utilized and will provide you with all the vitamin A that you need. Beta-Carotene and vitamin A are both measured in international units or IU. Ten thousand to 25,000 IU capsules or tablets should provide you with all your body's normal requirements.

•B-complex 50 or 100: Provides 50 or 100 milligrams (or micrograms) of all members of the B-vitamin complex. Unless you have need for a particular B vitamin, you will probably want to take B vitamins in this way instead of separately. Because B vitamins occur naturally as a complex, each element is important for efficient utilization. Moreover, when certain B vitamins are taken in isolation, deficiencies in other members of the complex can occur. Because most commercial brands of B vitamins or multivitamin supplements do contain not the full complex, you should take particular care in selecting a brand. You need to look for a supplement that not only contains the numbered B vitamins, but also the other elements such as folic acid, PABA, pantothenic acid, and inositol.

• Vitamin C comes in a variety of forms and formulas. If you are sick or under inordinate stress (either because of work conditions or emotional trauma), then you will want to take particularly large doses of vitamin C in conjunction with a good B-complex. Both of these vitamins are water soluble. They dissolve in water and do not accumulate within the body. Consequently, you may want to take smaller dosages of supplements a few times a day so that tissues become saturated, rather than in one large dose. When the body is under extreme stress, enormous quantities of vitamin C can be utilized and absorbed by the body. A powdered form dissolved in water or juice is best utilized when large dosages are needed. When tissues are saturated, you reach the maximum point of utilization called bowel tolerance, which can result in mild diarrhea.

Powdered vitamin C should always be taken in the ascorbate or buffered form like calcium or sodium ascorbate. Ascorbic acid is too acidic for large doses. Taken in pill or capsule, a C-complex containing rutin, hesperidin, and bioflavinoids is better utilized than ascorbate or ascorbic acid alone. In addition to aiding absorption, these members of the C complex play significant roles in the production of collagen, the connective matter that holds our skin, tissues, and organs together.

• Vitamin E is one vitamin that needs to be of a natural source, rather than synthetic, in order to be effective. Natural sources can be determined by their names. When the source of the vitamin is listed as d-alpha tocopherol or d-alpha tocopherol acetate, then it is a natural derivative. Any form containing the l- prefix (dl-alpha or l-alpha) is synthetic. Some vitamins will not list their source and will merely say "pure vitamin E" or "alpha tocopherol." Since it is more expensive to produce the natural–source vitamin E, those not indicating source are probably syntheti,c since the natural–source vitamins can be sold for more money. Vitamin E also appears in a complex, alpha tocopherol being the most active element. For maximum absorption, good vitamin lines will carry vitamin E in the form of mixed tocopherols including not only the alpha- but also beta- and gamma-tocopherols.

Like vitamin A, vitamin E is sold in international units measuring its biological activity. A capsule of 400 IU is sufficient for most people. In addition to being a potent antioxidant (or free radical scavenger), vitamin E is essential for the production of many of the body's hormones. Adele Davis, one of the foremost nutritionists of our times, believed that vitamin E, taken in conjunction with adequate amounts of the B vitamin PABA, helped to prevent hair from graying, and in some cases actually reversed the graying process.

• The essential minerals can be obtained through a multimineral formula. Minerals should always be in the chelated form, otherwise they will not be properly absorbed. But in the end, you must do your own reading and become informed to the point where you can make your own decisions as to what best protects your health.

In addition to supplements, the foods that you eat can greatly increase or decrease both your exposure to environmental toxins and your resistance to them. If, for example, you eat a regular diet of fast food, or even a diet of commercially produced food, you can be sure that you are getting a good dose of pesticides, food additives, salt, sugar, rancid fats, dyes, colorings, and preservatives. If you eat meat, then antibiotics and synthetic hormones will also be coursing through your system. Consequently, food can have a twofold effect on your health and well-being. First, if you eat commercially produced food, you are adding toxins to your system, toxins that tax the body's immune and cleansing systems. Second, in addition to the burden, you also are not getting the nutrients required to feed and stimulate immune and eliminative functions.

Food production today is geared toward profit maximization. Profits are maximized by foods that can stay on shelves for a long time and, be easily transported and marketed. Things like canned, frozen, packaged, and fast food fit the bill in this regard. But only at the expense of vitamins and other nutrients that we need to maintain health.

A shift toward a more natural, fresh, and localized food market is the first step toward a healthier food supply—and also a healthier body. In areas like San Francisco, southern California and New York, we already see these shifts taking place on a large scale. A Texas-based chain of gourmet health–food stores called the Whole Foods Company, provides a good example of what a well-educated consumer base can expect retailers to begin offering once the demand is great enough. Expanding on the concept of the health–food store, Whole Foods offers a wide variety of fresh–baked breads, natural cosmetic lines, an extensive organic–produce section, but also adds things like wines and a large selection of gourmet cheeses.

Also in Texas, the state's progressive commissioner of agriculture, Jim Hightower, is encouraging local growers to form cooperatives and aggressively promote their produce to local retailers and restaurants. In 1989, CBS's *60 Minutes* broadcast a segment on the success of this program. The same thing is happening throughout the country. Farmers' markets are gaining customers in many areas; restaurants wanting the freshest and tastiest produce are finding that local production best fits their needs. Public pressure to decrease pesticide residues is also leading farmers to begin combining traditional farming techniques with innovative methods of pest control.

Consumers have it within their power to influence the type of food they see in their markets. Food is a major contributor to our health and our ability to fight disease. We cannot eliminate all toxins and pollutants from our environment, but we can eat so as to minimize the damage these substances do to our bodies. In order to do this, foods must be fresh as vitamins, minerals, and enzymes diminish over time. Many are destroyed through cooking, canning, and freezing. Produce that is harvested green never attains its full nutritional value. Whole grains have plenty of natural fiber, aiding in digesting and decreasing the risk of colon and prostate cancer. These foods are also low in fat and calories.

Overweight and obese people are particularly at risk from pollution, as toxins accumulate in fat. Dieting can be a dangerous thing for people who have eaten a lifelong diet of commercially produced foods and have had high exposure to other environmental toxins. Anyone who is dieting should avoid fad and quick weight-loss programs. Not only are these programs usually not effective in the long run, but if you do begin to break down body fat, toxins can be released in large doses.

Exercise is important, not only for weight loss, but also for optimal health in general. Aerobic exercise for 20 to 30 minutes, four to five days a week, will raise your "set point" or the rate at which you metabolize food. Many people complain about gaining weight as a normal consequence of aging. It is true that our metabolisms do slow down as we age, but this can be almost fully compensated for by a regular exercise program. If you have a good physician, you may want to consult him or her about a good exercise program. For many people, a good swift walk will be all you need. If you have an especially busy schedule and you "just never have the time to exercise," you will also want to figure out ways to include exercise within your daily routine. For example, you may be able to walk to work, or walk during your lunch hour. You can also take stairs instead of elevators in many instances, and you can certainly walk up escalator steps.

Exercise, of course, helps to control weight, but it also plays a role in many other important body functions. Aerobic exercise helps vitamins to circulate and penetrate deep into cells. It increases circulation and elimination of toxins. Sustained aerobic activity will also cause the release of chemicals within the brain called endorphins. The primary role of these substances is to increase our pain threshold, but their secondary effect is to leave us with a sense of well-being, producing what athletes often refer to as "runner's high." Exercise is one of the best antidotes for emotional and mental stress, which many are beginning to recognize as one of the major contributors to disease and aging.

How We Determine Our Environment

Let's look first at agriculture, since it is the nexus between our internal and our external environment. We have discussed how the food we eat can determine our internal health and our ability to resist external pollution. But agriculture is also a source of many of the environmental problems facing us today. Consequently, by making a shift in the foods we eat, and demanding that they be produced in ways that are more sustainable, we simultaneously begin to solve some of our other pressing environmental concerns.

Right now we are at a crossroads. A substantial segment of the public is demanding a safer and more nutritious food supply. Scientific groups are indicating a need for change away from the production agriculture that started in the 1950s toward a more sustainable system. But the status quo, bolstered by the promises of biotechnology, means megaprofits for the agricultural establishment. Food production as it stands now is almost unfathomable in its complexity. What we eat is determined by pharmaceutical, chemical, and pesticide manufacturers, fast–food chains, advertising agencies, and oil companies. This massive financial endeavor is supported and promoted by the federal government, as well as many universities and research scientists. It is not just the quality of our food supply that has been adversely affected by this high-yield, high-profit system of agriculture and food production. America's insatiable appetite for beef is one of the primary reasons behind the destruction of the Amazon rain forests. Pesticides are contaminating drinking–water supplies across the country, while agricultural runoff causes eutrophication of our lakes and oceans. Our present system of subsidies, price supports, and other agricultural programs encourages inefficiencies, wastes taxpayer dollars,

and skews market forces that would otherwise direct farm resources into channels more in line with consumer demands. A
federal program paying farmers *not* to produce on their land in
1989 alone cost the country $2 billion.

Every one of us is in a position to vote with our consumer dollars
not only for the kind of food that we want, but also to end practices
that are causing environmental destruction. If you do not support
the destruction of the Amazon forests, then you can decrease you
consumption of beef. Let your food retailers know that you will not
purchase beef or beef products that are encouraging environmental devastation. By boycotting pesticide-laden produce, you
get safer food, and also let farmers know that if they want to sell
their products, they are going to have to come up with alternative
means of production. Support farmers who use techniques like
integrated pest management and organic farming. All of us need
to let food producers, retailers, and politicians know that irradiated food is not acceptable. It has no benefit for the consumer, and
merely adds to our nuclear–waste–disposal problems while at the
same time encouraging nuclear–weapons production.

Parents need to take charge of educating their children on good
food and proper eating habits, thereby retrieving the power they
have renounced to advertisers. Soft drinks, fast food, sugary cereals, and the host of other commonly advertised products are poison for all of us, but because of children's low body weight and pro-
portionally greater consumption of these foods, their harm is even
greater. Many health–food stores now carry a wide variety of
snack foods that are much lower in salt and sugar, have no additives, and are made.with organically grown grains. A few examples that you may want to try with your children include fruit–
juice–sweetened cookies, jams, cereals, juices, and chips. Children's tastes develop at an early age. Commercial products are so
concentrated in sugar, salt, and flavor enhancers that once children
become accustomed to them, they are unable to enjoy foods that
are not laden with these additives.

Turning now to some of our other environmental problems, there
is plenty that you can do in your daily life to contribute to their
solutions as well. For too long, we have believed that we all existed
in separate compartments; that we could do whatever we wanted,
within certain limits, and there would be no effect on our neighbors. If we are to begin to clean up our environment, a fundamental
shift is required. An important starting point is to recognize that
we live in an interrelated and interdependent world. Everything
that we do has an impact on the rest of the world. Environmental

problems cannot be viewed in isolation. Nowhere is this clearer than in the depletion of the ozone layer by substances like chlorofluorocarbons. The aerosol that you use today not only affects everyone on the planet now, but will continue to have that effect for many years to come.

This applies to virtually all environmental issues and is a clear illustration of the power of the individual when it comes to solving these problems. With respect to waste disposal, the only long-term solution lies in each of us beginning to cut down on our consumption of nondisposable items (plastics, polystyrene, etc.) and beginning to actively participate in recycling. Many stores now offer consumers the alternative betweenpaper or plastic shopping bags. Each time you opt for the paper, you contribute to decreasing waste at its source. You can save containers and buy in bulk. Many health– food stores now offer this alternative for grains, legumes, and nuts.

There are many things that each of us can do concerning air pollution. Recycling not only cuts down on waste, but also diminishes the amount of energy required to manufacture things like paper and plastics. Decreasing other sources of energy consumption will also benefit air quality. Depending on your lifestyle, you can walk, cycle, ride-share, or take public transportation to work. Buying more energy–efficient automobiles is also important. Most of us can significantly cut down energy consumption by using both heat and air–conditioning only when absolutely necessary. The important thing is for each person to make an effort in whatever way they can. This will depend upon where you live and what your lifestyle is. People in California, for instance, need to be particularly diligent in encouraging the construction and use of public transportation. In the summer months, those of us with yards can decrease energy consumption by hanging clothes on the line instead of using the dryer. One of our greatest causes of air pollution (as well as noise and visual pollution) comes from the fleet of diesel-powered tractor-trailers that continually boom across the nation's highways. All of us can encourage local production and distribution of food so that wasteful practices of transporting foods across country are decreased substantially. We can also let our legislators know that rail transportation for long hauls needs to be resurrected.

Many forms of pollution stem from the manufacture of consumer goods—PCBs in refrigerators, Superfund sites, and groundwater pollution from toxic–chemical dumping, dioxin in paper manufacturing. In this regard as well, there is plenty that each of us can do.

First, we can opt for consumer products that are not made with harmful chemicals. We can buy clothing and furniture that is not treated with formaldehyde and toxic dyes. By recycling, we can decrease the amount of chemicals needed to produce goods. In opting for purer foods, we can diminish the demand for pesticides. Although advertising has most Americans convinced that they need an arsenal of (usually toxic) substances to clean their homes, in many instances good old soap and water or solutions made up of baking soda or vinegar will do the job just as well. There are also a number of household products now on the market that are much more environmentally benign. Shifting to these safer products will not only improve the quality of your indoor air, but will also decrease the amount of toxic chemicals that go down your drain to pollute our oceans and waters.

While all of these suggestions may sound insignificant in and of themselves, we need to remember that it has been the cumulative effect of negative actions that has created many of our problems to begin with. As negative acts upon negative acts have caused a spiral of environmental destruction, so too will our positive and concerted actions function to reverse this damage.

Politics and the Environment

Thus far, I have avoided discussing the role of politics in cleaning up our environment. This is because many people have got into the habit of believing that the government and our representatives are going to handle situations, and consequently, there is no need for individual action. But this ignores one of the fundamental principles upon which this country is founded, namely that the power lies in the people. It is time for us to take back that power.

We have seen throughout this book that positive and constructive action with respect to environmental issues has, for the most part, come from individual citizens and public–interest groups. Laws have been passed, recycling programs instituted, and products banned from the market primarily at the behest of citizen action. This is one illustration of the power held by the people.

We also hold the power of the vote. Through our votes, we elect representatives, and through our votes we can remove them and change the political structure. We need to realize that environmental issues require strong leadership, people dedicated to serving the public interests in the long-term rather than the short-term objectives of powerful industry lobbies. When enough of us become sufficiently adamant, we will begin to see some long– overdue restructuring.

Most of our environmental problems can be tackled head-on by citizen action. Because of its inextricable ties to the military and long-standing government policies, nuclear pollution is a somewhat different situation. Many Americans are already having a significant influence on a grass-roots level by expressing their opposition to nuclear–power plants. Despite continued support from Washington and nonstop lobbying by the nuclear–power industry, widespread opposition by citizens has succeeded in thwarting construction of new plants, and in some cases has forced the closure of existing ones. But civilian power plants contribute to only a fraction of the nuclear waste now requiring a safe burial ground.

We have no idea how much money it is going to cost us to dispose of existing stocks of radioactive materials from power plants and weapons facilities. Some fear that we may never be able to safely dispose of these materials no matter how much money we spend. We can, however, stop generating more nuclear waste as of today. What that means is no more nuclear–weapons production, no more nuclear–power plants, and no food irradiation. The Pentagon and the weapons–manufacturing lobby has done a good job to date in convincing us that we cannot just stop making weapons. That would jeopardize national security, put the entire nation at risk, leave us open to invasion—in short, such a proposal is hopelessly naive. But apart from the destabilizing effects of the arms race, the simple truth is that our ever–mounting piles of nuclear waste are by far a greater threat to the health and well-being of the nation.

The American public needs to take a long hard look at this country's arms policy. We need to look at our legacy of waste, inefficiencies, and the total lack of accountability that has typified our weapons-making endeavors. The Department of Energy now admits that it deliberately released radioactive materials into populated areas, knowing that the materials could cause health damage. Now, after years of neglecting waste disposal and safety precautions, the DOE wants billions of dollars added to its budget to effectuate cleanups the department should have been handling all along.

One thing that we all need to understand is that whenever this billion is needed to clean up Superfund sites, or those billions are needed to dispose of the DOE's waste, we are talking about taxpayer dollars. Many politicians today insist that we need to raise taxes even more to balance the budget or to pay for costly new programs. I believe that precisely the contrary is true. It is this enormous till, this already massive flow of taxpayer dollars that gives government agencies a sense of separation from the reality of issues and allows for massive boondoggles like the nuclear–waste dilemma we see today.

Taxpayer dollars are all too often called upon to foot the bill for industry. It is our money that is paying to clean up the Superfund sites, not that of the polluting industries. When chemicals like DDT or EDB are believed to be carcinogenic, millions of taxpayer dollars go into the research and testing of these products, often for periods as long as 10 to 15 years. Even when the original test data are found to be defective, the manufacturer is not called upon to reconduct these tests or to prove safety. Many times government research that has cost taxpayers thousands,if not millions or billions, of dollars is turned over by researchers to inure to their own benefit. We have seen this with nuclear technology,* and now with biotechnology.

Conflicts of interest, greed, and profiteering are not solved by targeting a few scapegoats. On a political level, citizens need to demand a greater accountability from their representatives. In terms of tax dollars, if politicians say that they need more money, before they come to the people with a tax increase, let them first verify that the tax dollars they already have are being properly and efficiently used. We need to let politicians know that it is not acceptable to even talk about a tax increase when scandal after scandal in our government agencies (HUD, DOE, the Pentagon, etc.) reveals waste and mismanagement on a massive scale. The same holds true for antiquated and inefficient programs like the USDA's price supports and subsidies.† The problem is not a lack of funds, but a lack of leadership, oversight, and management by our politicians.

In the same way that we can cast our vote in political elections, we can also vote for corporate responsibility with our consumer dollars. Concerning the oil spill by the *Exxon Valdez*, for example, we

---

*Weapons research and development technology was turned over to private utilities for nuclear–power plants. Proponents of nuclear power often cite its low cost as one factor in its favor, but rarely are costs of research and development considered. With respect to food irradiation, one of the government scientists pioneering its technology left the government to start his own food irradiation company.

†This is not to say that tax dollars should not be spent on farm–assistance programs. However, current USDA programs need serious reevaluation. Many of these programs are purely wasteful, with no redeeming social purpose. Two billion dollars to keep land out of production is an example of such a program. USDA policy also encourages overproduction and inefficiency.

need to let politicians know that we do not support laws that give Exxon tax deductions for the costs of the cleanup. Some conscientious Americans immediately sent their Exxon credit cards back to the company as a protest of the way that Exxon had handled the affair. But that was only a handful of people. In the future we cannot wait for government to change laws that we find unjust. More of us, all of us, need to let companies like Exxon know that even if the laws do not mandate corporate responsibility, we—the company's customers—do.

A Prescription for Total Well-Being

Here are some specific ideas for ways that you can help keep your environment clean. Remember that the more you can fit into your lifestyle, the more you will be contributing to a cleaner environment.

Personal Grooming
•Use hand and body soap without any artificial scent.
•Use soap instead of shaving cream (which often contains ammonia and ethanol, both of which are dangerous).
•Make shampoo out of one cup liquid Castile soap mixed with a half-cup distilled water and 1/4 cup olive or avocado oil.
•Instead of commercial hair conditioner, use sesame or corn oil before you shampoo.
•Mix peppermint extract with baking soda for a refreshing toothpaste.
•Make deodorant out of equal parts baking soda and cornstarch .
•Use cornstarch instead of talcum powder (asbestos-free).
•Sesame oil is an effective sun block. It also contains vitamin E which is a good moisturizer.
•Use vegetable oil instead of baby oil to smooth skin or remove makeup

Energy Consciousness

•Turn up or down the thermostat to conserve energy before leaving your house.

•Try to be as fuel–efficient as you can on your way to work. Walk if you can, or if you can't, use a bicycle. Lobby local government for bike lanes on roads.

•Instead of driving alone, use mass transit. If you must drive, car pool with friends or coworkers.

At Work

•Stand back from the photocopier and always make sure the lid is closed when using it.

•Use a monitor filter (also called a glare screen) which reduces ELF and VLF radiation given off by a computer's screen.

•Use compressed air guns to remove dust from computer keyboards regularly to eliminate the need for toxic, volatile chemical solvents.

•Try to get out of the office for a least half an hour daily at lunchtime, even a few minutes for a walk out in nonrecirculated air increases energy levels, relieves fatigue, and refreshes both mind and body.

At Work, continued

•If your job involves contact with solvents, toxic chemicals, or much dust, make sure to wear appropriate protective equipment such as rubber gloves, goggles, and gas masks.

•Offices go through a lot of paper. Encourage your employer to conserve paper use, and reuse old memos and announcements for scrap and notepaper. Also, encourage your employer to recycle paper.

At the Market on the Way Home

•Bring your own bags to the supermarket— plastic and paper bags can be used several times, and the fewer you throw out, the more we cut down on pollution.

•Try to buy organic or naturally grown foods. There are now ranchers who raise livestock and poultry without use of hormones, antibiotics, or pesticide–treated feeds. If your grocer does not carry organically grown meat, request that he stock it.

•Try to avoid processed foods. When you can't, read labels carefully and stay away from products with preservatives, artificial coloring, sweeteners, and stabilizers.

•Use paper plates instead of plastic–coated or polystyrene ones for outdoor meals and picnics.

At the Market on the Way Home

•Buy domestically grown produce—imported foods contain higher levels of pesticides and preservatives.

•Wash all fruits and vegetables to get rid of surface pesticides. Peel root vegetables or any coated with wax (e.g., cucumbers). Remove outer leaves of lettuce and cabbage.

•When barbecuing, use regular charcoal instead of the self–lighting kind. Light coals with paper and wood kindling instead of lighter fluid.

•Reuse aluminum foil.

•Use waxed paper instead of plastic wrap.

•Be aware of the containers your food comes in. If you buy a premium brand of orange juice, for example, does it come in a plastic jug or cardboard carton like milk? If it is a carton, is it biodegradable or foil–lined?

•Reuse plastic containers that food comes in instead of throwing them out. They have many uses around the house, from holding nuts and bolts to storing leftovers.

•If you have a garden, compost organic waste.

•Separate paper goods, cans, and bottles from other waste and recycle.

At the Market on the Way Home , continued

•Use cloth diapers instead of disposables. Most large cities have diaper services that pick up and deliver. If you're set on disposables, used biodegradable ones. (*TenderCare* are 97% biodegradable and chemical-free. Available through mail order, 5555 East Seventy-First Street, Suite 8300, Tulsa, OK 74136, (800) 344-6379.)

•Don't rely on disposable batteries. Use rechargeable ones instead. Batteries are made of toxic heavy metals that corrode and seep into landfills, contaminating water supplies with cadmium, mercury, and lead.

Replace Hazardous Household Products

•Use mineral or castor oil to lubricate squeaky hinges, latches, and locks.

•To decrease household odors, grow plants to absorb excess carbon dioxide and boil herbs and spices on the stove to freshen the air.

•Use butter to remove grease from hands before washing.

•Soften paintbrushes with hot vinegar.

•Use Borax as an all–purpose cleaner—mix with water for spray cleaner.

Replace Hazardous Household Products, continued

•Clean up spills in the oven and on the stovetop immediately—
that way they don't get baked on and there's no need to use
powerful, toxic oven cleaners.

•Use baking soda as scouring powder in kitchen and bathroom—mix with equal parts soap for an effective abrasive.

•Mix three tablespoons vinegar with one quart water to make
spray glass cleaner.

•Use washing soda mixed with soap instead of commercial
laundry detergent. Wash clothes once with pure washing soda
to eliminate all residue from detergents before employing the
mixture to prevent yellowing.

•Add one cup of vinegar to the final rinse cycle for a good fabric
softener.

•Mix two tablespoons of cornstarch to a pint of water to create
a natural spray starch. Shake before spraying.

•For pest problems, instead of using insect sprays containing
neurotoxins such as chlorpyrifos, purchase roach motels.

•Use boric acid as a safe way to get rid of ants and cockroaches.
To prevent them from returning, plant mint or onions (insect–
repelling plants) around the house.

Replace Hazardous Household Products, continued

•In the garden, use Safer Insecticidal Soap for Fruits and Vegetables, a nontoxic insecticide based on naturally occurring fatty acids. It kills most garden pests, and will not harm beneficial bugs.

•When landscaping, consider alternative ground cover instead of a grass lawn. This conserves water in addition to reducing the necessity for chemical fertilizers.

When Making Improvements to Your House

•Water–based (latex) house paints release fewer toxins than oil –based ones. Check the labels for volatile organic compound (VOC) numbers, the lower the better.

•Check the energy efficiency rating (EER) on appliances and buy appliances that are as efficient as possible—this saves money on utility bills in the long run, and the less energy we use, the less fossil fuel we burn.

•Before undertaking any remodeling, check to see whether the affected area contains asbestos or urea-formaldehyde insulation. If it does, these must be professionally removed to eliminate potential health hazards.

•Radon is a naturally occurring gas that is a by-product of the breakdown of uranium found in almost all soil. It is usually discovered in basements. Prolonged exposure has been found to result in lung cancer. Radon is easily detectable

When Making Improvements to Your House, continued

•(Radon continued), through a number of moderately priced home detection kits. If you find radon is collecting in your home, the easiest remedy is to increase ventilation throughout the house. Since the gas dissipates when mixed with air, the situation can frequently be remedied by installing fans and vents to the outside, especially in your basement.

•Make sure your house is well insulated. Insulation reduces fuel consumption and costs. Weatherstrip windows—they should be as airtight as walls.

•Flow restrictors in shower heads reduce the amount of hot water used, which also conserves energy.

ACCESS GUIDE

<u>Whom to Write in Congress</u>

Below is a listing of senators and representatives currently in office with interests in the environment. They can be reached at the following addresses.

U.S. Senate
Washington, DC 20510

U.S. House of Representatives
Washington, DC 20515

Sen. Morris Udall (D-Arizona)—chairman of House Committee on the Interior
Sen. George Mitchell (D-Maine)—champion of the Clean Air Act
Sen. Al Gore (D-Tenn)—first member of Congress to fight for legislation to reduce the effects of greenhouse, including imposing controls on CFCs
Sen. Timothy Wirth (D-Colo)—member of Senate Energy and Natural Resources Committee
Sen. John Chafee (R-RI)—member of Environment and Public Works Committee
Rep. Bruce Vento (D-Minn)—advocate of national parks, wilderness areas, and temperate and tropical forests
Rep. Henry Waxman (D-Cal)—advocate of legislation for clean drinking water and toxic contamination

The Environmental Protection Agency

You can also contact your regional office of the EPA with questions or problems. Their toll-free hotline in Washington, D.C. is (800) 424-4000. In addition, the EPA can be reached at the following addresses:

EPA Region 1
JFK Federal Building
Boston, MA 02203
(617) 565-3234

EPA Region 2
26 Federal Plaza
New York, NY 10278
(212) 264-4418

EPA Region 3
841 Chestnut Street
Philadelphia, PA   19107
(215) 597-4084

EPA Region 4
345 Courtland Street, NE
Atlanta, GA   30365
(404) 347-2904

EPA Region 5
230 South Dearborn Street
Chicago, IL 60604
(312) 886-6165

EPA Region 6
1445 Ross Avenue
Dallas, TX 75202
(214) 655-7208

EPA Region 7
726 Minnesota Avenue
Kansas City, KS 66101
(913) 236-2893

EPA Region 8
999 18th Street
One Denver Place, Suite 1300
Denver, CO   80202
(303) 293-1648

EPA Region 9
215 Freemont Street
San Francisco, CA   94105
(415) 974-8378

EPA Region 10
1200 Sixth Avenue
Seattle, WA 98101
(206 442-7660

Multi-Issue Groups

Friends of the Earth Foundation
530 Seventh Street, SE
Washington, DC 20003
(202) 654-4312
Issues including soil conservation, toxic wastes, acid rain, water quality, tropical forests, and pesticides. Encourages education, sponsors research, publishes annual reports and educational materials.

Greenpeace
1611 Connecticut Avenue, NW
Washington, DC 20009
(202) 462-1177
Issues including acid rain, ocean dumping, toxic wastes, and other environmental concerns. Organizes demonstrations and rallies encouraging protection of species, publishes fact sheets, and sponsors regional environmental libraries. Newsletter, *Greenpeace Examiner*.

Citizens for a Better Environment
33 East Congress
Suite 523
Chicago, IL 60605
(312) 939-1530
Educational group with branches in five states that works through litigation and lobbying. Canvasses and works on the grass-roots level to change people's buying and recycling habits. Researches manufacturing facilities (results published in the *Toxic Air Report*). Focuses on solid waste, recycling, air and water pollution, and low-level radioactive waste. Publishes a quarterly magazine, *Environmental Review*, and several fact sheets.

Center for Holistic Resource Management
P.O. Box 7128
Albuquerque, NM 87194
(505) 242-9272
Focuses on resource management with practical human, financial, and biological goals. Works on an international level and publishes a quarterly newsletter as a membership service. Textbook and workbook on holistic resource management.

Concern, Inc.
1794 Columbia Road, NW
Washington, DC 20009
(202) 238-8160
Publishes the *Community Action Guide* on water, farmland, pesticides, waste, and household waste; distributed internationally to schools, governments, businesses, and individuals.

International Ecology Society
1471 Barclay Street
Saint Paul, MN   55106
(612) 774-4971
Animal and environmental protection group that lobbies Congress and publishes an infrequent newsletter and fliers.

Institute for Resource and Security Studies
27 Ellsworth Avenue
Cambridge, MA   02139
(617) 491-5177
Focuses on nuclear and conventional disarmament, environmental protection, and sustainable human activity. Publishes reports and several books as well as researching for publications. Works on education through the media and other organizations (such as Greenpeace).

South West Research Information Center
P.O. Box 4524
Albuquerque, NM   87106
(505) 262-1862
Focuses on mining issues, solid waste, nuclear waste, community right-to-know, groundwater, oil and gas production, and resource development issues. Provides education and technical assistance nationally to community organizations, provides legislative analysis, lobbying, and acts as expert witness in court cases. Quarterly magazine, *The Workbook*, and frequent staff papers.

Worldwatch Institute
F1776 Massachusetts Avenue, NW
Washington, DC 20036
(202) 452-1999
Environmental-research group that focuses on ozone, food production, alternative energy, and population. Publications include a bimonthly magazine and papers, such as the annual *State of the World Report* every February, which is frequently used by researchers and Congress.

Citizen/Labor Energy Coalition
225 West Ohio
Suite 250
Chicago, IL 60610
(202) 875-5153
Grass-roots consumer organization that deals with health care, toxic waste, renewable fuels, and insurance reform. Works through lobbying and research, and uses media impact for legislative changes. Quarterly newsletter.

U.S. Public Interest Research Group
215 Pennsylvania Avenue, SE
Washington, DC 20003
(202) 546-9707
Environmental consumer group concerned with ozone, the greenhouse effect, pesticides, and the Clean Air Act. Research and advocacy group that lobbies for better legislation. Citizen outreach campaign to educate the public and gain member support. Quarterly newsletter, *Citizens Agenda*.

Inform
381 Park Avenue South
New York, NY 10016
(212) 689-4040
Research and educational group that specializes in solid-waste management, alternative vehicle fuels, and irrigation in the west. Considered to be on the cutting edge of knowledge in these fields, and their data is frequently drawn upon by policymakers on the federal, state, and municipal levels. Quarterly newsletter.

Multi-Issue Publications

Ecosphere
Forum International: International Ecosystems University
91 Gregory Lane, No. 2
Pleasant Hill, CA 94523
(415) 946-1500
Bimonthly tabloid describing theory and practice of "ecosystemic whole-world-oriented, transdisciplinary, value-based education, research, and action programs." Also, book reviews. $18 yearly.

State Of The States
Fund for Renewable Energy and the Environment (FREE)
1001 Connecticut Avenue, NW
Suite 638
Washington, DC 20036
(202) 466-6880
Annual report providing a critical look at environmental progress
at the state level in the following areas: air-pollution reduction, soil conservation, groundwater protection, hazardous-waste management, solid
waste and recycling, and renewable energy and conservation. Price: $10
per copy.

Educational Organizations
Society for Educational Reconstruction
c/o T. M. Thomas, Dept. of Education
University of Bridgeport,
Bridgeport, CT   06602
(203) 624-3687
Formed by a group of university teachers, a national organization that
works for society and peace through the way they teach in the classroom.
Occasionally involved in legislative efforts. Publishes a newsletter three
times a year.

CIEP Fund
68 Harrison Avenue
Boston, MA 02111
(617) 426-4375
Focused on national environmental issues. Works to place college students and recent graduates in paid, short-term environmental jobs in private industry, government, and nonprofit organizations. Fosters a nonad-
versarial relationship between governmental, private, and environmental
groups by building environmental professionalism. Conferences, workshops, and seminars. Publishes a quarterly newsletter, *Connections*, and
books, *Complete Guide to Environmental Careers* and *Becoming an Environmental Professional: 1990*.

Coolidge Center for Environmental Leadership
1675 Massachusets Avenue, Suite 4
Cambridge, MA 02138
(617) 864-5085
International environment and development group that addresses multiple issues by organizing educational programs for Third World countries. Seminars and weekend retreats taught by experts work toward the

empowerment of local people for sustainable development. Publishes a newsletter three times a year and a calendar of events around the Boston area.

Computer Networking

Experimental Cities
P.O. Box 731
Pacific Palisades, CA   90272
(213) 276-0686
Founded in 1972, built Earthlab. Publishes a newsletter (soon to go on-line). Developing computer conferencing and networks and holds ongoing meetings linking nonprofit groups. Working on democratic participation over computers and increasing efficiency of networking for groups. Works with international Green movement, World Citizens Assembly, teachers, and Sane/freeze.
Political Action Groups

League of Conservation Voters
2000 L Street, NW   #804
Washington, DC 20036
(202) 785-8683
Electoral arm of the environmental movement, existing only to endorse and donate money to candidates who are pro-environment. Creates the *National Environmental Score Card,* which rates congressional voting records on the environment. Publishes a national election report and profiles presidential races.

Political Publications

Environmental Action
Environmental Action
1525 New Hampshire Avenue, NW
Washington, DC 20036
(202) 745-4870
Bimonthly magazine covering the environmental movement and its goals and policies for lay, academic, activist, and political audiences. Also book reviews, econotes, toxic roundup. Price: included in membership dues.

Sierra Club Legal Defense Fund—In Brief
Sierra Club Legal Defense Fund
2044 Fillmore Street
San Francisco, CA 94115
(415) 567-6100
Quarterly newsletter on environmental law focusing on membership activities. Free.

Products and Services

Seventh Generation
Department 60M89
10 Farrell Street
South Burlington, VT 05403
A mail-order business that offers environmentally safe household products, such as nontoxic cleaners and solar-powered accessories, and environmental books.

Pennsylvania Resources Council
P.O. Box 88
Media, PA 19063
(215) 565-9131
Sells a two-dollar *Environmental Shopping Guide* along with a 20-page booklet that lists environmentally friendly products.

Toxic and Natural: How to Avoid Dangerous Everyday Products and Buy or Make Safe Ones
c/o Debra Lynn Dodd
Box 1506
Mill Valley, CA 94942
This book/catalog is an excellent source for safe household products.

TOXNET
Specialized Information Services Division
National Library of Medicine
8600 Rockville Pike
Bethesda, MD 20894
(301) 496-6531
A 24-hour database system set up especially for occasional users, TOXNET leads callers step-by-step through the process of cleaning up toxic spills, small or large. Available through TELENET, TYMNET, or CompuServe telecommunications networks.

Food Safety

Public Voice for Food and Health Policy
1001 Connecticut Avenue, NW
Suite 522
Washington, DC 20036
(202) 659-5930
Consumer research, education, and advocacy group that promotes the public interest in decision making on food and health issues that include: food labeling, fish inspection, pesticide regulation, biotechnology in the food supply, and the nutritional status of the rural poor. Publications include: monthly letter to consumer organizations, quarterly *Action Alert* to members, and a subscription service for decision makers that includes testimonies and legislative information.

Center for Health Action
P.O. Box 270 Forest Park Station
Springfield, MA 01108
(413)   782-2115
National group focusing on stopping water fluoridation. Advocates personal health care and freedom of health choices, works for educating the public through press releases, and publicizes bills in Congress. Quarterly newsletter, pamphlets, and books.

Safe Water Coalition
150 Woodland Avenue
San Anselmo, CA 94960
(415) 453-0158
Information source that focuses on stopping water fluoridation. John Lee, M.D., works with the coalition to research fluoridation and has written a paper on the Gilbert Syndrome, a jaundicelike disease that can be turned on and off with fluoride. Publishes the *National Fluoridation News* quarterly.

Organic Farms
10726B Tucker Street
Beltsville, MD 20705
(800) 222-6244
One of the nation's largest distributors of organically grown foods, also provides nationwide listing of organic restaurants.

Farming Organizations
Universal Proutist Farmers Federation
1354 Montague, NW
Washington, DC 20011
(202) 882-8804
Concerned with small organic farmers, promotes government financing
and training of farmers. Based on the Prout philosophy of shared resources for guaranteed basic human needs. Provides information and
sponsors a resource center on credit unions, cooperatives, organic farming, and legal information. Newsletter, *Farming the Future.*

New Alchemy Institute
2376 Hatchville RD
East Falmouth, MA 02536
(617) 564-6301
Researches, educates, and runs internship programs. Deals with cover
cropping, composting, greenhouse management, and integrated pest
management, especially in the Northeast. Sustainable design integrating
the above areas. Quarterly newsletter.

International Alliance for Sustainable Agriculture
Newman Center University of Minnesota
1701 University Avenue, Room 202
Minneapolis, MN 55414
(612) 331-1099
Promotes systems of agriculture that value social justice, economy, and
ecology. Primary issues: pest control, pesticides, sustainable agriculture,
and farm practices. Sponsors educational conferences, exchanges, and
tours. Resource center, referral service, quarterly newsletter, *Manna*, and
various other publications.

Natural Organic Farmer's Association
RFP #2
Barre, MA 01005
(413) 247-9264
Grass-roots organization of consumers, gardeners, and farmers that promotes cleaner food and healthier environment. Supports methods of
farming and gardening that can continue with future generations and that
show respect for soil, water, and air. Promotes political action supporting
stable, local agricultural practices and publishes *Natural Farmer* three
times a year. Certifies organic farms, promotes interstate reciprocity and
creation of common standards for certification.

Farm Labor Organizing Committee
507 South Claire Street
Toledo, OH 43602
(419) 243-3456
Nonprofit union for migrant farm workers with contacts to Campbell Soup, Vlasic Pickles, Heinz, and others.

North American Farm Alliance
Box 2502
Ames, IA 50010
(515) 232-1008
Formed in 1983, operates across the nation to preserve small family farming through education. Also lobbies state legislators and helped to form the national *Save the family Farm Coalition* based in Washington. Influences agricultural policy, advocates sustainable agriculture, published *North American Farmer* monthly, sells T-shirts and hats that say "FARMS, NOT ARMS."

Organic Foods Production Association of North America
226 East Second Street
Winona, MN 55987
(507) 452-6332
Multi-issue trade association made of many suborganizations. Works by educating retailers, wholesalers, and consumers, and publishes a book on certification guidelines and a yearly magazine, *Organic Foods.* Syracuse University professor Kate Clancy leads the subcommittee on processing guidelines, which helps to write laws on organic food. Its members have been involved in laws in 26 states.

Society for Agricultural Training and Integrated Voluntary Activities
Green
Route 2, Box 242W
Viola, WI 54664
(608) 625-2217
Focuses on self-sufficient gardening and education through hands-on experience and volunteer labor, trading room and board for work on farms. Quarterly newsletter and a listing of organic farms in the Midwest.

National Save the Family Farm Coalition
80 F Street, NW
Washington, DC 20001
(202) 737-2215
Membership organization formed in 1986 that lobbies for the economics of small farms and family farms, and advocates sustainable agriculture. Supports reduction from large farms to small farms as well as the 1990 farm bill, which removes the barriers to environmentally sound farming. Most members are local family members. Publishes quarterly newsletter, $25 a year.

Alternative Farming Systems Information Center
National Agricultural Library
Room 111
Beltsville, MD 20705
(301) 344-3724
Provides information and advice on how to farm and garden organically.

National Coalition Against the Misuse of Pesticides
530 Seventh Street, SE
Washington, DC   20003
Provides extensive information on toxic-free pest control for indoors or outdoors.
Farming Publications

New Alchemy Quarterly
New Alchemy Institute
237 Hatchville, Road
East Falmouth, MA 02536
(617) 564-6301
Quarterly newsletter covering current research and practical applications of sustainable technologies for providing food, energy, shelter, and landscape design. Each issue devoted to central theme. Price: Included in membership dues; two dollars per copy to nonmembers.

Energy

Critical Mass Energy Project
215 Pennsylvania Avenue, SE
Washington, DC 20003
(202) 546-4996
National advocacy group founded in 1974 by Ralph Nader that deals with energy conservation and renewable resources. Lobbies Congress, files litigation, and monitors federal agencies. Works with citizens' groups and publishes studies, interviews, and papers for media release in addition to bimonthly *Critical Mass Energy* .

Alternative Energy Resources Organization
324 Fuller— C4
Helena, MT 59601
(406) 443-7272
Primarily interested in sustainable agriculture and renewable energy sources. Offers workshops and training seminars, provides research and public education.

The National Food Energy Council
409 Vandiver West, Suite 202
Columbia, MO 65202
(314) 875-7155

Rocky Mountain Institute
1739 Snowmass Creek Road
Snowmass, CO 81654
Provides information on superinsulated homes, water conservation, and resource-efficient living.

Conservation and Renewable Energy Inquiry and Referral Service
Box 8900
Silver Spring, MD 20907
(800) 523-2929
Provides extensive information on energy conservation.
Energy Publications

Alternative Energy: A Guide to Free Information to Educators
Fund for Renewable Energy and the Environment
1001 Connecticut Avenue, NW
Suite 638
Washington, DC 20036
(202) 745-4874

Powerline
Environmental Action Foundation
1525 New Hampshire Avenue, NW
Washington, DC 20036
(202) 745-4874
Bimonthly tabloid covering consumer issues on energy and utilities; includes legislative and regulatory reports. Also reviews books and other information resources. Price: $15 for individuals and grass-roots groups, $10 for low-income and senior citizens.

Nuclear Power

Land Educational Associated Foundation
3368 Oak Avenue
Stevens Point, WI 54481
(715) 344-6158
Researches nuclear power, waste, and weapons, and works through education. Reprints and redistributes books, publishes quarterly newsletter.

Mobilization for Survival
45 John Street, Room 811
New York, NY 10038
(212) 385-2222
Currently organizing local action in communities against Department of Energy regarding military weapons facilities. Pursuing research on the more general environmental impacts of all types of military facilities. Quarterly newsletter.

Environmental Coalition on Nuclear Power
433 Orlando Avenue
State College, PA 16803
(814) 237-3900
Founded in 1970. Focuses on radioactive waste management and licensing of Pennsylvania nuclear-power plants. Educates via conferences on health effects of nuclear power, addresses regulatory issues and legislation, and gives assistance in writing legislation. Newsletter.

National Campaign for Radioactive Waste Safety
P.O. Box 4524
105 Stanford, SE
Albuquerque, NM 87106
(505) 262-1862
Information source concerned with government handling of radioactive waste. Publishes *The Workbook*.

Radioactive Waste Campaign
625 Broadway, 2nd Floor
New York, NY 10012-2611
(212) 437-7390
Environmental advocacy and public-interest organization that promotes greater public awareness of the danger to human health and the biosphere from the production, transport, and storage of radioactive waste. Publishes books, fact sheets, slide shows, and a quarterly newspaper on radioactive-waste issues, *The Waste Paper*.

Citizens Energy Council
77 Homewood Avenue
Allendale, NJ 07401
(201) 327-3914
Oldest opposition group to nuclear power and nuclear weapons, has members in 40 states and foreign countries. Sponsors workshops and debates on both the national and state levels and will arrange for speakers and debates. Claims partial credit for preventing the construction of 80 nuclear-power plants. Sponsors the World Congress on Radiation Hazards at United Nations on November first. Newsletter, *Radiation Perils.*

Citizen's Call
P.O. Box 1722
Cedar City, UT   84720
(801) 268-0186
Helps radiation victims and people who are affected by radioactive materials, focusing on people downwind of the Nevada test site, but works on a national level as well. Networks with medical, military, and mining groups, consults with Congress as expert witness and has started a hospice organization for victims. Intermittent newsletter.

Nuclear Issue Publications

Natural Power Newsletter
Natural Power
5420 Mayfield Road
Cleveland, OH 44124
(216) 442-5600

Monthly newsletter advocating the replacement of nuclear-power plants with natural, nonpolluting, alternativ sources of energy. Price: five dollars per issue for nonmembers.

<u>Antipollution Organizations</u>

National Clean Air Coalition
1400 16th Street, NW
Washington, DC 20036
(202) 797-5436
Coalition of 36 national organizations working to strengthen the Clean Air Act through intense lobbying, grass-roots efforts, and media presence. Primarily a national group, but also works with 20 million local grass-roots activists to put forth legislative proposals.

National Water Alliance
1225 First Street, NW
Suite 300
Washington, DC 20005
(202) 646-0917 ext. 204
Nonadvocacy group that forms a consensus which is used by congress-men, senators, corporate leaders, environmental leaders, and academic fields. Publishes a book, *Water: Today's Agenda*.

Clean Water Action
317 Pennsylvania Avenue, SE
Washington, DC 20003
(202) 547-1196
National lobby group that visits toxic dumping sites and makes reports on legislative situations. Quarterly newsletter.

Water Information Network
P.O. Box 4525
Albuquerque, NM 87106
(505) 262-1862
Network of individuals and organizations in and around New Mexico that links grass-roots groups around water quality. Introduces and sup-ports bills and holds regional meetings and conferences. Quarterly news-letter.

Center for Short Lived Phenomena
P.O. Box 199, Harvard Square Station
Cambridge, MA 02238
(617) 492-3310
Provides consulting and testimony for government and publishes a bi-weekly newsletter on oil spills and cleanup.

Environmental Action
1525 New Hampshire Avenue, NW
Washington, DC 20036
(202) 745-4870
National membership group that researches clean air, water, and solid waste, as well as lobbies for bills before Congress. Two bimonthly maga-zines for subscribers.

Earth Regeneration Society
1442A Walnut Street #57A
Berkeley, CA 94709
(415) 525-7723
waste. Supported by members. Publishes a local environmental directory and the bimonthly *Upstate Environment*. Free library.

The Acid Rain Foundation
1410 Varsity Drive
Raleigh, NC 27606
(919) 737-3520
Works with teachers to educate the public. Has various publications that
deal with all issues as they relate to acid rain.

Alliance for Clean Energy
1901 North Ft. Meyer Drive, 12th floor
Rosslyn, VA 22209
(703) 841-1781
Coalition of low-sulphur coal producers, users, and transporters interested in the acid-rain problem. Uses legislative action and initiates grass-
roots work to ensure the industry is represented.

American College of Ecology
Route 1, Box 269
Clayton, IN 45118
(317) 539-4653
Consists of a variety of state groups and provides speakers, participants
for lobbying, and legislative monitoring concerned with soil, water, and
air pollution.

Pollution Publications
Environmental Review
Citizens for a Better Government
33 East Congress, Suite 523
Chicago, IL 60605
(312) 939-1530
Quarterly journal informing members and other readers on the public-
health effects of pollution and the preventive measures reducing pollution by toxic substances; includes updates of the organization's activities,
research reports, and an annual report in the January/February issue.
Price: Included in membership dues. $20 per year for nonmembers.

Association of Local Air Pollution Control Officials— Washington Update
Association of Local Air Pollution Control Officials
444 North Capitol Street, NW
Suite 306
Washington, DC 20001
(202) 624-7864
Newsletter covering congressional and EPA activities, current issues re-
lated to air pollution, and association news. Also a calendar of events and
Federal Register notice summaries. Price: included in membership dues.

Recycling and Waste

Ecology Center
1403 Addison Street
Berkeley, CA 94702
(415) 548-2220
Main focus is public education and environmental alternatives. Offers recycling workshops as well as library and bookstore. Newsletter and various other publications.

American Forestry Association
1516 P Street, NW
Washington, DC 20005
(202) 667-3300
Primary issues include education of public on proper use of natural resources, including forests, wildlife, and water. Sponsors wilderness trips, provides access to landowners, and promotes legislative alternatives. *American Forests* magazine and other publications.

Renewable Natural Resources Foundation
5430 Grosvenor Lane
Bethesda, MD 20814
(301) 493-9101
Coalition of natural-resource groups concerned with future air, land, and water supplies. Supports organizations via educational services, symposiums, and research. Also operates the renewable natural-resources center and publishes *Renewable Resources* journal quarterly.

Citizens Clearinghouse for Hazardous Wastes
P.O. Box 926
Arlington, VA 22216
(703) 276-7070
Grass-roots environmental-crisis center that helps local community groups by providing technical and organizational assistance to a network of more than 5,000 grass-roots community groups. Publishes 50 handbooks relating to environmental hazards and two quarterly newsletters.

Environmental Action Coalition
625   Broadway
New York, NY 10012
(212) 677-1601
Focuses primarily on solid-waste management, but also deals with water quality and water conservation, conducts research, and provides educa-

tional curriculum for all levels of schools. Also implements local recycling and waste reduction programs and gives free regional workshops. Publishes the newsletter *Cycle* and a *Curriculum Publications* list.

The National Campaign Against Toxic Hazards
37 Temple Place, 4th Floor
Boston, MA 02111
(617) 482-1477
National group that has worked on the Superfund and the Stratospheric Ozone Projection Act (HR-2699) to ban ozone destroying products. Offers technical support (chemists, industrial hygienists, an environmental lab) as well as a legal team to advise. Has an influence on capacity insurance, a 20 year hazardous waste plan to reduce incinerators, and publishes *The Consumers Guide to Ozone Protections* as well as a quarterly magazine. Also helps the victims of toxic poisoning and has worked with Greenpeace, Clean Water Action, and the PIRGS

Waste Watch
P.O. Box 39185
Washington, DC 20016
(213) 475-1684
Informal national network that works issue by issue, but focuses on pollution prevention and recycling. Lobbies infrequently and distributes information, also publishes a book, *Waste Watcher*.

Clean Sites, Inc.
1199 North Fairfax Street
Alexandria, VA 22314
(703) 683-8522
Mediates between parties at cleanup sites, has an advisory staff of lawyers and engineers, and works nationally with the EPA Publishes a report, *Making Superfund Work*, which was presented to the president, and a newsletter, *Clean Sites.*

Community Environmental Council
930 Miramante Drive
Santa Barbara, CA 93109
(805) 963-0583
Models recycling programs with schools, businesses, and restaurants. Some issues include: solid and hazardous-waste management, public policy, land use, and environmental review. Conducts policy seminars, publishes research and policy papers, advises local government and businesses and has a demonstration organic garden. Publishes a quarterly newsletter, brochures, and pamphlets.

Center for Plastics Recycling Research
Rutgers University
Building 3529, Busch Campus
Piscataway, NJ 08855
Provides information and manuals on collecting and sorting disposed plastic products, and operating a plastics-recycling plant.

Society of the Plastics Industry
1275 K Street, NW
Suite   400
Washington, DC 20005
(202) 371-5200
Provides comprehensive listing of companies that recycle plastics; waste and recycling publications

Bioenergy   Update
National Wood Energy Association
P.O. Box 498
Pepperell, MA 01463
(617) 433-5674
Bimonthly tabloid focusing on the improvement of American wood resources, especially the use of wood for energy. Offers updates on development of wood as a renewable energy source through the ruse of responsible management practices. Also a calendar of events, legislative updates, and research news. Price: included in membership dues, $25 a year for nonmembers.

Gildea Review
Community Environmental Council
930 Miramante Drive
Santa Barbara, CA 93109
(805) 963-0583
Quarterly membership activities newsletter covering local, state, and national environmental issues with an emphasis on waste management, land use, recycling, water conservation, and public decision making. Also book reviews. Price: included in membership dues, five dollars per issue for nonmembers.

Keep America Beautiful- Vision
Keep America Beautiful (KAB)
Mill River Plaza
Nine West Broad Street
Stamford, CT   06902
(203) 323-8987
Quarterly newsletter covering litter prevention, voluntary recycling, community improvement programs, and activities of KAB. Calendar of events and new members. Also publishes an annual report on KAB programs of. Price: first copy free.

NAOSMM Newsline
National Association of Scientific Materials Managers
Chemistry Department, University of New Orleans
New Orleans, LA 70148
(504) 286-6324
Quarterly newsletter covering waste disposal, the relationship between exposure to chemicals and cancer, the U.S. Occupational Safety and Health Administration, and association news. Price: included in membership dues.

Sources And Further Reading

During the preparation of this book, events affecting the environment unfolded on a daily basis. The author is indebted to the science, health and business reporters of the *The New York Times* and *The Wall Street Journal* for their detailed coverage of the changing scene. These include: Phillip Shabecoff, Matthew L. Wald, Keith Schneider, Jane Brody, Amy Dockser Marcus, Ken Wells and Marilyn Chase, A. Paszter and R. Taylor, H. Myers, and Sonia Nazario.

The information included in this book represents the most recent available at press time.

Where We Live and Work

New Age Journal, April 1986, p. 46

Indoor Pollution News, June 1, 1989

U.S. Environmental Protection Agency, *Report to Congress on Indoor Air Quality*, Washington, D C

Rogers, Sherry, "The Many Guises of Mold Allergy," *Bestways*, January 1985, p. 44

Rogers, Sherry, "Tired or Toxic," *The Human Ecologist*, No. 36, p. 9

The Healthy House Catalog, available from The Healthy House, 4115 Bridge Ave., Cleveland, OH 44113 (800) 222-9348 ($15 plus $2.50 postage and handling)

The Inside Story: A Guide to Indoor Air Quality, U. S. Environmental Protection Agency, Office of Air and Radiation, Washington, DC 20460 (free)

The Greenhouse Effect

Cooke, Robert, "Life in a Greenhouse," *New York Newsday*, September 13, 1988, Part III, p. 1

Sinclair, Lani, *Changing Climate: A Guide to the Greenhouse Effect* , World Resources Institute, 1709 New York Ave., NW, Washington, DC 20006 (free)

Cooling the Greenhouse: Vital First Steps to Combat Global Warming, Natural Resources Defense Council, 1350 New York Ave., NW, Washington, DC 20005 ($5)

Policy Options for Stabilizing Global Climate: Draft Report to Congress, U.S. Environmental Protection Agency, PM 221, 401 M Street, SW, Washington, DC 20460

The Potential Effects of Global Warming on the United States: Draft Report to Congress (Executive Summary), U.S. Environmental Protection Agency, PM 221, 401 M Street, SW, Washington, DC 20460

Ozone Depletion

Shell, Ellen Ruppel, "Watch This Space," *Omni*, August 1987

Washington Post Wire Service, "97% Drop in Ozone Recorded Over Antarctica," *The Star Ledger*, October 28, 1987, p. 21

Natural Resources Defense Council, "Ozone Depletion Worsens," *Newsline,* p. 1
Russell, Dick, "The Endless Simmer," *In These Times,* January 11, 1989, p. 10
"Can We Repair the Sky?" reprint from *Consumer Reports,* P.O. Box CS2010-A, Mount Vernon, NY 10553
Protecting the Ozone Layer: What You Can Do, Environmental Defense Fund, 257 Park Avenue South, New York, NY 10010 ($2)
Saving the Ozone Layer: A Citizen Action Guide, Natural Resources Defense Council, 122 East 42nd Street, New York, NY 10168
Stones in a Glass House—CFC's and Ozone Depletion, Investor Responsibility Research Center, 1755 Massachusetts Avenue, NW, Washington, DC 20036 ($35)

Acid Rain
National Audubon Society, "Monthly Report of pH Readings," *Citizen's Acid Rain Monitoring Network,* January 1988
Air Pollution, Acid Rain and the Future of Forests, Worldwatch Institute, 1776 Massachusetts Avenue, Washington, DC 20036 ($4)
Trends in the Quality of the Nation's Air, U.S. Environmental Protection Agency, Office of Public Affairs, 87-019, Washington, DC 20460

Water
"Cleanup of a Pure River," *East/West,* June 1987, p. 13
Brown, Larry J. and Deborah Allen, "Toxic Waste and Citizen Action," *Science for the People,* July/August 1983, p. 6
Barton, K., "Fish Cancer," *The Environment,* November, 1983, p. 24
Kuchenberg, "Measuring the Health of the Ecosystem," *Environment,* March, 1985, p. 32
Johnson, Tim, "The Poisoned Turtle," *Daybreak,* Summer 1988, p. 25
Remba, Zev, "Beyond Water Wasteland," *Clean Water Action News,* Summer 1987
Burke, William K., "An Effluent Community Worries...." *In These Times,* April 12, 1989, p. 8
Thompson, Dick, "The Greening of the U.S.S.R.," *Time,* January 2, 1989, p. 68
McCarthy, Jane, "Fishy Mystery of Dead Sea on N.J. Shore," *New York Post,* August 29, 1988, p. 14
Thompson, Dick, "Stains on the White Continent," *Time,* February 20, 1989, p. 77
"Today's Robber Barrons Despoil the Environment," *In These Times,* April 12, 1989, p. 14
Lemonick, Michael D., "The Two Alaskas," *Time,* April 17, 1989, p. 56
King, Jonathan, "Troubled Water," *Rodale's Practical Homeowner,* January 1987,
Langone, John, "A Stinking Mess," *Time,* January 2, 1989, p. 47
Lefferts, Lisa and Stephen Schmidt, "Water: Safe To Swallow?," *Nutrition Action Health Letter,* November 1988, p. 5
Sibbison, J., "The Battle to Ban Fluorides," *Bestways,* December 1982, p. 60
Geiser, Ken and Gerry Waneck, "PCBs and Warren County," *Science for the People,* July/August 1983, p. 13
Sibbison, J., "Ground Water Contamination," *Bestways,* October 1985
Steppacher, Lee and Tara Gallagher, "Natural Resources," *Environment,* May 1988,
The Chesapeake Bay Foundation Homeowners Series: Water Conservation, Household Hazardous Waste, Detergents, the Chesapeake Bay Foundation, 162 Prince George Street, Annapolis, MD 21401
A Citizen's Guide to River Conservation, The Conservation Foundation, 1250 24th Street, NW, Washington, DC 20037

Danger On Tap, The Government's Failure to Enforce the Federal Safe Drinking Water Act, National Wildlife Foundation, 1400 16th Street, NW, Washington, DC 20036
Drinking Water—A Community Action Guide, Concern, Inc., 1794 Columbia Road, NW, Washington, DC 20009 ($3)
Gabler, Raymond, *Is Your Water Safe to Drink?*, Consumer Reports Books ($16)

Modern Farming
Lehrman, Sally, "In A Flooded Market, Farms Awash in Debt," *San Francisco Examiner*, March 17, 1986, p. A1
The American Farmland Trust, *National Survey*
Ludlow, Lynn, "The Grapes Of Wrath— 1986 Edition," *San Francisco Examiner*, March 17, 1986, p. A4
Bean, Kenneth, "The Farm Crisis," *Utne Reader*, Nov/Dec 1987, p. 46
Conrat, M. and R., *The American Farm*, Houghton Mifflin, 1977
Wood, Wilbur, "Distorted Vision," (Pacific News Services) reproduced in *Utne Reader*, Nov/Dec 1987, p. 48
"Ag Chemicals; Economics Incentives vs. Environmental Goals," *Farm and Dairy*, December 22, 1988, p. 9
Chowka, Peter, "Down on The Farm," *New Age*, October 1982, p. 42
Lappe, F. and J. Collins, *Food First*, Houghton Mifflin, 1977
"New Tomatoes Put the Bite on Bugs," *Yuba-Sutter Appeal-Democrat*
"Scientists See Designer Animals," *Yuba-Sutter Appeal-Democrat*
Berry, Wendell, *The Unsettling of America*, Sierra Club, 1977
"Rex Oberhelman: $27,000 from Five Organic Acres," *Mother Earth News*, March/April 1986, p. 17
Ambrose, Jon, "Natural Pest Control," *Natural Food and Farming*, May 1989, p. 16
Gips, Terry, *Breaking The Pesticide Habit*, International Alliance for Sustainable Agriculture, 1987

Pesticides
Laseter, J., et al., "Chlorinated Hydrocarbon Pesticides in Environmentally Sensitive Patients," *Clinical Ecology*, Vol. 11, No. 1, Fall 1983, p. 3
"National Health Council Report," *An Assessment of Health Risk of Seven Pesticides Used for Termite Control: 9*, Washington, D.C., National Academy Press, 1982
Johnson, K., "Equity In Hazard Management," *Environment*, November 1982, p. 29
Goldfarb, T. and D. Wartenberg, "Fighting Pesticides on Long Island," *Science for the People*, Jan/Feb 1983, p. 18
Eggington, J., "Temik Troubles Move South," *Audubon*, May 1983
McGuire, Rick, "Dioxin Pollution Called Pervasive," *Medical Tribune*, December 24, 1986, p. 5
Van Strum, Carol, "A Bitter Fog," *East/West Journal*, July 1983, p. 48
Hathaway, Janet, "Eating Wisely Gets Harder All the Time," *Los Angeles Times*, October 8, 1987
Sibbison, J., "Why We're Still Eating Pesticides," *Bestways*, December 1982, p. 30
Natural Resources Defense Council, *Pesticides in Food*, March 15, 1984
Kaplan, Sheila, "The Food Chain Gang," *Common Cause Magazine*, Sept./Oct. 1987
Jones, Pamela, "Chavez and Policy Group Inflate Pesticide Poisonings," *American Council on Science and Health News and Views*, May/June 1987, p. 1
Carlson, Margaret, "Do You Dare to Eat a Peach?," *Time*, March 27, 1989, p. 24
Defusing the Toxic Threat: Controlling Pesticides and Industrial Waste, Worldwatch Institute, 1776 Massachusetts Avenue, NW, Washington, DC 20036 ($4)
Yepsen, Roger Jr., *The Encyclopedia of Natural Insect and Disease Control*, Rodale Press, Emmaus, PA

Food Additives
McNair, James, *Cheese*, Chronicle Books, San Francisco, 1986
Toufexis, Anastasia, "Dining With Invisible Danger," *Time*, March 27, 1989, p. 28
"Coke Isn't It," *East/West Journal*, November, 1983, p. 9
"The Unending Attack On The American Diet," *Science Digest*, August 1985, p. 15
"Coconuts in Your Coffee," *American Health*, Jan./Feb. 1987
"Crohn's Linked to Margarine and Choc Creams," *Medical Tribune*, April 17, 1985,
Brown, Michael H., "Fast Foods Are Hazardous To Your Health," *Science Digest*, April 1986, p. 31
Kreisman, Richard, "McDonald's Nutrition Ads Test," *Advertising Age*, March 9, 1981
"Fast Food Nutrition Unveiled," *East/West*, July 1986, p. 13
Carroll, James R., "Carcinogenic Dyes Still Out, Panel Reports," *Tucson*, June 5, 1985
Turner, James S., *Chemical Feast: Report on the FDA*, (Ralph Nader Study Group Reports), Grossman, 1970

Food Irradiation
"Comments on the Food Marketing Institute's *Report on Food Irradiation*," Food and Water, Inc., Denville, NJ, July 15, 1988
Bhaskaram, C. and G. Sadasivan, "Effects of Feeding Irradiated Wheat To Malnourished Children," *American Journal Of Clinical Nutrition*, Vol. 28, 1975, p. 130
Terry, Ken, "Why Is DOE for Food Irradiation," *The Nation*, February 7, 1987, p. 143

Nuclear Waste
Hohenemser, Christoph and Ortwin Renn, "Chernobyl's Other Legacy," *Environment*, April 1988, p. 5
Greenwald, J., "Deadly Meltdown," *Time*, May 12, 1986, P. 38
Jeffrey, J.W., "The Collapse of Nuclear Economics," *The Ecologist*, Jan./Feb. 1988, p. 9
Bunyard, Peter, "The Myth of France's Cheap Nuclear Electricity," *Ecology*, Jan./Feb. 1988, p. 4
Miller, Mark, "Not So Bad After All?," *Newsweek*, July 25, 1988, p. 65
Miller, Jean, "Major Radium Dump Found in New York City," *RWC Waste Paper*, Fall 1988, p. 3
Danger Downwind: A Report on the Release of Billions of Pounds of Toxic Air Pollutants, National Wildlife Federation, 1400 16th Street, NW, Washington, DC 20036
"Superfund: Looking Back, Looking Ahead," *EPA Journal*, Jan/Feb 1987
Electromagnetic Pollution
"Pentagon's Electromagnetic Tests Halted," *St. Petersburg Times*, May 15, 1988
"Engineer Stands By Figure," *Register-Guard*, Eugene, OR, March 28, 1978, p. 3A
"FCC Discounts Oregon Radio Wave Peril," *Los Angeles Herald Examiner*, March 29, 1978, p. A8

The Toxic Time bomb
Krattenmaker, Tom, "Warning Sent To Shore Polluters," *Daily Record*, Morris County, NJ, April 23, 1989, p. B12
"Plastics Tax Slated For Senate Vote," *Daily Record*, Morris County, NJ, April 23, 1989

Magnuson, E., "A Problemthat Cannot Be Buried," *Time*, October 14, 1985, p. 76
"Right Train, Wrong Track: New Report on E.P.A.'s Mismanagement of Superfund," *NRDC Newsline*, July/August 1988, p. 1
"Recycling Garbage: Hard but Promising," *Inform*, May-June 1985, p. 3
Sombke, Laurence, "Cut The Garbage," *USA Weekend*, April 21, 1989, p. 4
Stoler, P., "Turning to New Technologies," *Time*, October 14, 1985, p. 90
Coming Full Circle: Successful Recycling Today, Environmental Defense Fund, Inc., 257 Park Avenue South, New York, NY 10010 ($20)
Mining Urban Wastes: The Potential for Recycling, Worldwatch Institute, 1776 Massachusetts Avenue, NW, Washington, DC 20046 ($4)
Waste: Choices for Communities, Concern, Inc. 1794 Columbia Road, NW, Washington, DC 20009 ($3)

General Interest
Carson, Rachel, *Silent Spring*, Houghton Mifflin Co., Boston, 1962
Environmental Pollution: A Long-Term Perspective, World Resources Institute, 1709 New York Avenue, NW, Washington, DC 20006
Environmental Progress and Challenges: EPA's Update, U.S. Environmental Protection Agency, Public Information Center, PM211B, 401 M Street, NW, Washington, DC 20460
Personal Action Guide for the Earth, Friends of the United Nations—Transmissions Project, 730 Arizona Avenue, Santa Monica, CA 90401
"The Planet Strikes Back! 21st Environmental Quality Index," *National Wildlife Magazine*, Feb./March 1989, Educational Publications, National Wildlife Federation, 8925 Leesburg Pike, Vienna, VA 22184
Shopping for a Better World: A Quick and Easy Guide to Socially Responsible Supermarket Shopping, Council on Economic Priorities, 30 Irving Place, New York, NY 10003 ($4.95)

Index

ABOUT THE AUTHOR

GARY NULL hosts a series of nationally syndicated one-minute health-and-nutrition spots, "Total Health," as well as a television show carried in 97 cities across the country. His radio talk show, *The Gary Null Show*, is carried nationwide by the American Radio Network.

He holds a Ph.D. in human nutrition and public health science is a nutritionist, consumer advocate, and environmental activist who has written over 20 books as well as numerous articles for national magazines on health, diet and nutrition, and lectures on these topics around the country. In addition, he has been involved with the environmental movement for over 20 years. His most recent books include *Gary Null's Complete Guide to Healing Your Body Naturally* and *The Complete Guide to Health and Nutrition*. He is a national champion racewalker who lives in New York City.